advancing learning, changing lives

Salters-Nuffield Advanced Biology
for Edexcel AS Biology

D0495250

STUDENTS' BOOK

A PEARSON COMPANY

Pearson Education Limited
Edinburgh Gate
Harlow
Essex
CM20 2JE
United Kingdom

and Associated Companies throughout the world

© University of York Science Education Group 2008

First published 2005
Published as trial edition 2002
This edition published 2008

ISBN 978-1-4058-9607-8

The publisher's policy is to use paper manufactured from sustainable forests.

SNAB project editor: Anne Scott
Edited by Kate Redmond
Designed and illustrated by Pantek Arts, Maidstone, Kent
Picture research by Charlotte Lipmann
Index by Laurence Errington

Printed and bound by Graficas Estella, Bilboa, Spain

Contents

Contributors

Many people from schools, colleges, universities, industries and the professions have contributed to the Salters-Nuffield Advanced Biology project. They include the following.

Central team

Angela Hall	Nuffield Curriculum Centre
Michael Reiss	Institute of Education, University of London
Anne Scott	University of York Science Education Group
Sarah Codrington	Nuffield Curriculum Centre

Authors

Angela Hall	Nuffield Curriculum Centre	Cathy Rowell	Bootham School, York
Sue Howarth	Tettenhall College	Anne Scott	University of York Science Education Group
Nick Owens		Nicola Wilberforce	Esher College
Michael Reiss	Institute of Education, University of London		

Acknowledgements

We would also like to thank the following for their advice and assistance.

Teachers, technicians and students at schools and colleges running the Salters-Nuffield Advanced Biology course.

Steve Hall	King Edward VI School, Southampton	Professor Eve Roman	University of York
Liz Hodgson	Greenhead College	Sandra Wilmott	University of York Science Education Group
Professor Robin Millar	University of York		

Sponsors

We are grateful for sponsorship from The Salters' Institute and the Nuffield Foundation who have continued to support the Salters-Nuffield Advanced Biology project after its initial development and have enabled the production of these materials.

Authors of the previous editions

This revised edition of the Salters-Nuffield Advanced Biology course materials draws heavily on the initial project development and the work of previous authors.

Glen Balmer	Watford Grammar School	Laurie Haynes	School of Biological Sciences, University of Bristol
Susan Barker	Institute of Education, University of Warwick	Paul Heppleston	
Martin Bridgeman	Stratton Upper School, Biggleswade, Bedfordshire	Liz Jackson	King James's School, Knaresborough
Alan Clamp	Ealing Tutorial College	Christine Knight	
Mark Colyer	Oxford College of Further Education	Pauline Lowrie	Sir John Deane's College, Northwich
Jon Duveen	City & Islington College, London	Peter Lillford	Department of Biology, University of York
Brian Ford	The Sixth Form College, Colchester	Jenny Owens	Rye St Antony School, Headington, Oxford
Richard Fosbery	The Skinners School, Tunbridge Wells	Nick Owens	Oundle School, Peterborough
Barbara Geatrell	The Burgate School, Fordingbridge, Hants	Jamie Shackleton	Cambridge Regional College
Ginny Hales	Cambridge Regional College	David Slingsby	Wakefield Girls High School
Steve Hall	King Edward VI School, Southampton	Mark Smith	Leeds Grammar School
Gill Hickman	Ringwood School	Jane Wilson	Coombe Dean School, Plymouth, Devon
Liz Hodgson	Greenhead College, Huddersfield	Mark Winterbottom	King Edward VI School, Bury St Edmunds

Advisory Committee for the initial development

Professor R McNeill Alexander FRS	University of Leeds
Dr Roger Barker	University of Cambridge
Dr Allan Baxter	GlaxoSmithKline
Professor Sir Tom Blundell FRS (Chair)	University of Cambridge
Professor Kay Davies CBE FRS	University of Oxford
Professor Sir John Krebs FRS	Food Standards Agency
Professor John Lawton FRS	Natural Environment Research Council
Professor Peter Lillford CBE	University of York
Dr Roger Lock	University of Birmingham
Professor Angela McFarlane	University of Bristol
Dr Alan Munro	University of Cambridge
Professor Lord Robert Winston	Imperial College of Science, Technology and Medicine

Please cite this publication as: Salters-Nuffield Advanced Biology AS Student book, Edexcel Pearson, London, 2008

About the course

Context-led study

Salters-Nuffield Advanced Biology (SNAB) is much more than just another A-level specification. It is a complete course with its own distinctive philosophy. The course is supported by a comprehensive set of teaching, learning and support materials which embrace a student centred approach. SNAB combines the key concepts underpinning biology today, combined with the opportunity to gain the wider skills that biologists now need.

A context-led approach

In the Salters-Nuffield Advanced Biology approach you study biology through real-life contexts. For example, most A-level biology courses start with cell biology or biochemistry. We don't. We start with an account of Mark, a 15-year-old who had a stroke, and Peter, an adult who had a heart attack. You study the biological principles needed to understand what happened to Mark and Peter. You then go on from the details of their cases to look at the factors that make it more likely that any of us will suffer from a stroke or heart attack. All four AS topics use this context-led approach; a storyline or contemporary issue is presented, and the relevant biological principles are introduced when required to aid understanding of the context.

Building knowledge through the course

In SNAB there is not, for example, a topic labelled 'biochemistry' containing everything you might need to know on carbohydrates, fats, nucleic acids and proteins. In SNAB you study the biochemistry of these large molecules bit by bit throughout the course when you need to know the relevant information for a particular topic. In this way information is presented in manageable chunks and builds on existing knowledge.

Activities as an integral part of the learning process

SNAB encourages an active approach to learning. Throughout this book you will find references to a wide variety of activities. Through these, you will learn both content and experimental techniques. In addition, you will develop a wide range of skills, including data analysis, critical evaluation of information, communication and collaborative work.

Within the electronic resources you will find animations on such things as the cardiac cycle and cell division. These animations are designed to help you understand the more difficult bits of biology. The support sections should be useful if you need help with biochemistry, mathematics, ICT, study skills, the examination or coursework.

SNAB and ethical debate

With rapid developments in biological science, we are faced with an increasing number of challenging decisions. For example, the rapid advances in gene technology present ethical dilemmas. Should embryonic stems cells be used in medicine? Which genes can be tested for in prenatal screening?

In SNAB you develop the ability to discuss and debate these types of biological issues. There is rarely a right or wrong answer; rather you learn to justify your own decisions using ethical frameworks.

Exams and coursework

Edexcel examines SNAB AS as the context-led approach within the Edexcel AS Biology specification. The Edexcel exams reward your ability to reason scientifically and to use what you have learned in new contexts, rather than merely being able to regurgitate huge amounts of information you have learnt off by heart. Most of the exam questions are structured ones, but you will also write extended coursework reports. We believe that this will be very useful for you if you go on to university or to any sort of job that requires you to be able to write reports. You can find out more about the coursework and examinations within the electronic resources and in the specification.

We feel that SNAB is the most exciting and up-to-date advanced biology course around. Whatever your interests are – whether you just want to do the AS course or go on to A2 and study a biological subject at University – we hope you enjoy the course.

Any questions?

If you have any questions or comments about the materials you can let us know via the website or write to us at:

The Salters-Nuffield Advanced Biology Project
Science Curriculum Centre
University of York
Heslington
York YO10 5DD

www.advancedbiology.org

How to use this book

There are a number of features in the student books that will help your learning and help you find your way around the course.

This AS book covers the four AS topics. These are shown in the contents list, which also shows you the page numbers for the main sections within each topic. There is an index at the back to help you find what you are looking for.

Main text

Key terms in the text are shown in **bold type**. These terms are defined in the interactive glossary that can be found on the software using the 'search glossary' feature.

There is an introduction at the start of each topic and this provides a guide to the sort of things you will be studying in the topic.

There is an '**Overview**' box on the first spread of each topic, so you know which biological principles will be covered.

> ### Overview of the biological principles covered in this topic
>
> This topic will introduce the concept of risks to health. You will study the relative sizes of risks and how these are assessed. You will consider how we view different risks – our perception of risk. You will also look at how health risks may be affected by lifestyle choices and how risk factors for disease are determined.

Occasionally in the topics there are also '**Key biological principle**' boxes where a fundamental biological principle is highlighted.

> ### Key biological principle: The effect of increase in size on surface area
>
> Living organisms have to exchange substances with their surroundings. For example, they take in oxygen and nutrients and get rid of waste materials such as carbon dioxide. In unicellular organisms the whole cell surface membrane is the exchange surface. Substances that diffuse into or out of a cell move down a **concentration** **gradient** (from a high to a low concentration). The gradients are maintained by the cell continuously using the substances absorbed and producing waste. For example, oxygen diffusing into a cell is used for respiration which produces carbon dioxide.

'**Did you know?**' boxes contain material that will not be examined, but we hope you will find it interesting.

> ### ? Did you know?
>
> #### One cause of male infertility
>
> For the human zygote to develop, the gamete nuclei have to fuse and a chemical from the sperm cytoplasm is required to activate the fertilised cell. This chemical is a protein called oscillin. It causes calcium ions to move in and out of stores in the cytoplasm of the ovum. These oscillations of calcium ion concentration trigger the zygote to begin developing into an embryo. Oscillin is concentrated in the first part of the sperm to attach to the ovum, and enters before the male nucleus in order to activate the ovum. It is thought that low levels of oscillin in sperm may be linked to male infertility, and this is a current area of research.

Questions

You will find two types of question in this book.

In-text questions occur now and again in the text. They are intended to help you to think carefully about what you have read and to aid your understanding. You can self-check using the answers provided at the back of the book.

> **Q3.5** How many possible combinations of maternal and paternal chromosomes could be found in the gametes of organisms with $2n = 8$, and organisms with $2n = 10$?

Crossing over

During the first meiotic division, homologous chromosomes come together as pairs and all four **chromatids** come into contact. At these contact points the chromatids break and rejoin, exchanging sections of DNA (see Figure 3.16). The point where the chromatids break is called a **chiasma** (plural **chiasmata**), and several of these often occur along the length of each pair of chromosomes, giving rise to a large amount of variation.

> ### ✔ Checkpoint
>
> **3.3** Produce a concept map or table which summarises how genetic variation is generated.

Boxes containing '**Checkpoint**' questions are found throughout the book. They give you summary-style tasks that build up some revision notes as you go through the student book.

Links to the online resources

'**Activity**' boxes show you which activities are associated with particular sections of the book. Activity sheets and any related animation can be accessed from the activity homepages found via 'topic resources' on the software. Activity sheets include such things as practicals, issues for debate and role plays. They can be printed out. Your teacher or lecturer will guide you on which activity to do and when. There may also be weblinks associated with the activity, giving hotlinks to other useful websites.

A final activity for each topic enables you to '**check your notes**' using the topic summary provided within the activity. The topic summary shows you what you need to have learned for your unit exam.

'**Weblink**' boxes give you useful websites to go and look at. They are provided on a dedicated 'weblink' page on the software under 'SNAB communications'.

'**Extension**' boxes refer you to extra information or activities available in the electronic resources. The extension sheets can be printed out. The material in them will not be examined.

'**Support**' boxes are provided now and again, where it is particularly useful for you to go to the student support provision within the electronic resources, e.g. biochemistry support. You will also be guided to the support in the electronic resources from the activity home pages, or you can go directly via 'student support'.

GCSE reviews and interactive GCSE review tests are provided to help you revise GCSE biology relevant to each AS topic.

At the end of each topic, as well as the '**check your notes**' activity for consolidation of each topic, there is an interactive '**Topic test**' box. This test will usually be set by your teacher / lecturer, and will help you to find out how much you have learned from the topic.

The key biological principle and all boxes linking to online resources are colour coded for each topic.

Activity

In **Activity 4.18** you can have a go at extracting fibres and then testing their tensile strength. **A4.18S**

Weblink

To find out more about captive breeding programmes visit the European Association of Zoos and Aquaria website.

Extension

You can find out how to calculate allele frequencies in **Extension 4.1. X4.01S**

Support

To remind yourself about hydrogen bonds, visit the Biochemistry support on the website.

Lifestyle, health and risk

Why a topic called Lifestyle, health and risk?

Congratulations on making it this far! Not everyone who started life's journey has been so lucky. In the UK only about 70% of conceptions lead to live births, and about 6 in every 1000 newborn babies do not survive their first year of life (Figure 1.1). After celebrating your first birthday there seem to be fewer dangers. Fewer than 2 in every 1000 children die between the ages of 1 and 14 years old. All in all, life *is* a risky business.

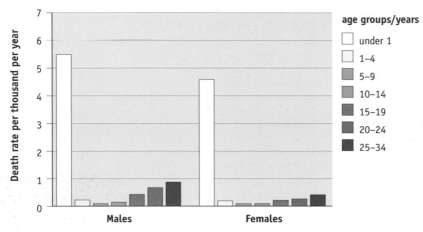

Figure 1.1 Death rates per 1000 population per year by age group and sex. Is life more risky for boys? *Source: England and Wales Office for National Statistics, 2004.*

In everything we do there is some risk. Normally we only think something is risky if there is the obvious potential for a harmful outcome. Snowboarding, parachute jumping and taking ecstasy are thought of as risky activities, but even crossing the road, jogging or sitting in the sun have risks, and many people take actions to reduce them (Figures 1.2 and 1.3).

Risks to health are often not as apparent as the risks facing someone making a parachute jump. People often do not realise they are at risk from a lifestyle choice they make. They underestimate the effect such choices might have on their health.

What we eat and drink, and the activities we take part in, all affect our health and well-being. Every day we make choices that may have short- and long-term consequences of which we may be only vaguely aware. What are the health risks we are subjecting ourselves to? Will a cooked breakfast set us up for the day or will it put us on course for heart disease? Does the 10-minute walk to work really make a difference to our health?

Cardiovascular disease is the biggest killer in the UK, with more than one in three people (37%) dying from diseases of the circulatory system. Does everyone have the same risk? Can we assess and reduce the risk to our health? Do we need to? Is our perception of risk at odds with reality?

Figure 1.2 Some activities are less obviously risky than others, but may still have hidden dangers.

In this topic you will read about Mark and Peter, who have kindly agreed to share their experiences of cardiovascular disease. The topic will introduce the underlying biological concepts that will help you understand how cardiovascular diseases develop, and the ways of reducing the risk of developing these diseases.

Overview of the biological principles covered in this topic

This topic will introduce the concept of risks to health. You will study the relative sizes of risks and how these are assessed. You will consider how we view different risks – our perception of risk. You will also look at how health risks may be affected by lifestyle choices and how risk factors for disease are determined.

Building on your GCSE knowledge of the circulatory system, you will study the heart and circulation and understand how these are affected by our choice of diet and activity.

You will look in some detail at the biochemistry of our food. This will give you a detailed understanding of some of the current thinking among doctors and other scientists about how our choice of foods can reduce the risks to our health.

Figure 1.3 A UK male aged 15 to 24 is over three times more likely to have a fatal accident than a female of the same age.

Review

Are you ready to tackle Topic 1 *Lifestyle, health and risk?*

Complete the GCSE review and GCSE review test before you start.

Mark's story

On 28 July 1995 something momentous happened that changed my life…

I was sitting in my bedroom playing on my computer when I started to feel dizzy with a slight headache. Standing, I lost all balance and was feeling very poorly. I think I can remember trying to get downstairs and into the kitchen before fainting. People say that unconscious people can still hear. I don't know if it's true but I can remember my dad phoning for a doctor and that was it. It took 5 minutes from me being an average 15-year-old to being in a coma.

Figure 1.4 Mark at 15.

I was rushed to Redditch Alexandra Hospital where they did some reaction tests on me. They asked my parents questions about my lifestyle (did I smoke, take drugs, etc.?). Failing to respond to any stimulus, I was transferred in an ambulance to Coventry Walsgrave Neurological Ward. Following CT and MRI scans on my brain it was concluded that I had suffered a stroke. My parents signed the consent form for me to have an operation lasting many hours. I was given about a 30% chance of survival.

They stopped the bleed by clipping the blood vessels that had burst with metal clips, and removing the excess blood with a vacuum. I was then transferred to the intensive care unit to see if I would recover. Within a couple of days I was conscious and day by day I regained my sight, hearing and movement (although walking and speech were still distorted).

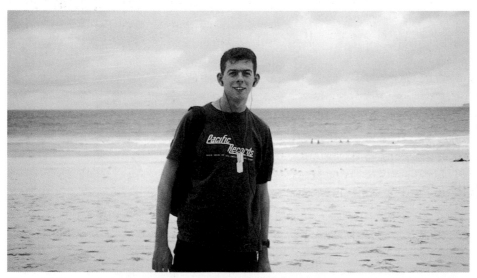

Figure 1.5 The experience is not stopping Mark living life to the full.

This is a true story. Mark had a stroke, one of the forms of cardiovascular disease. It is rare for someone as young as Mark to suffer a stroke. Why did it happen? Was he in a high-risk group?

Peter's story

I got the first indication of cardiovascular problems aged 23, when I was told that I had high blood pressure. I didn't really take much notice. My father had died at the age of 53 from a heart attack but as he was about four stone overweight, had a passion for fatty foods and smoked 60 full strength cigarettes a day, I didn't compare his condition to mine. I had a keen interest in sport, playing hockey and joining the athletics team at work. I was never overweight but I must admit that I probably drank too much at times and didn't bother too much about calories and cholesterol in food.

In 1981, I ran my first marathon at the age of 42 and subsequently did another five. All was going well I thought, until a routine medical showed my blood pressure reading to be 240 over 140. The doctor could not believe that I was still walking around, let alone running, and sent me straight to my GP. Since then I have always taken tablets for blood pressure and have also reviewed my diet.

I did continue running and completed the Great North Run at the age of 63. Thinking about doing the Great North Run again, I was running 8 miles a week and playing hockey. Then my eight-day holiday in Ireland became three days touring and twelve days in hospital.

At 2 o'clock in the morning I woke up with a terrific pain in my chest. I was sweating profusely and looking very pale. I had had a heart attack and within an hour I was in intensive care. At 5 am I had a second attack and the specialist inserted a temporary pacemaker to keep my heart rate up as it was dropping below 40.

Figure 1.6 Peter's active lifestyle did not prevent his heart attack but probably helped him to make a full recovery.

After five days in intensive care I was transferred to the general ward for recuperation. I was told that it was possible that, had I not looked after myself, I might have had a heart attack much earlier in life.

On returning home I had an angiogram and was told that I needed a triple bypass operation. I have to say it was not pleasant, but I had decided that it was necessary and I would cope with anything that happened if it would get me back to a decent lifestyle. Well, the operation, a quadruple bypass, was a success and after eight days I was back home.

This is a true story. Why did it happen to Peter, who seemed to be so active and healthy?

Activity

To find out what happened to Mark and Peter read their full stories in **Activity 1.1.**
A1.01S

1.1 What is cardiovascular disease?

Deaths from cardiovascular disease

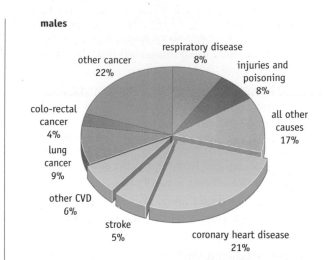

Figure 1.7 Premature deaths by cause in the UK in 2004 for females (left) and males (right). (Premature death is death under the age of 75 years.) One person dies of heart disease in the UK every 3 minutes. *Reproduced with the kind permission of the British Heart Foundation.*

Cardiovascular diseases (CVDs) are diseases of the heart and circulation. They are the main cause of death in the UK, accounting for over 200 000 deaths a year, and over 60 000 of these are premature deaths (Figure 1.7). More than one in three people in the UK die from cardiovascular diseases. The main forms of cardiovascular diseases are **coronary heart disease (CHD)** as experienced by Peter, and **stroke** as experienced by Mark.

About half of all deaths from cardiovascular diseases are from coronary heart disease and about a quarter are from stroke. Coronary heart disease is the most common cause of death in the UK. One in four men and one in five women die from the disease.

Weblink

To check out the most recent death rate figures for coronary heart disease see the National Statistics Office website.

Activity

Activity 1.2 demonstrates mass flow. **A1.02S**

Key biological principle: Why have a heart and circulation?

The heart and circulation have one primary purpose – to move substances around the body. In very small organisms, such as unicellular creatures, substances such as oxygen, carbon dioxide and digestive products are moved around the organism by diffusion. **Diffusion** is the movement of molecules or ions from a region of their high concentration to a region of their low concentration by relatively slow random movement of molecules.

Most complex multicellular organisms, however, are too large for diffusion to move substances around their bodies quickly enough. These animals usually have blood to carry vital substances around their bodies and a heart

to pump it instead of relying on diffusion. In other words, they have a circulatory system. Some animals have more than one heart – the humble earthworm, for instance, has five.

Open circulatory systems

In insects and some other animal groups, blood circulates in large open spaces. A simple heart pumps blood out into cavities surrounding the animal's organs. Substances can diffuse between the blood and cells. When the heart muscle relaxes, blood is drawn from the cavity back into the heart, through small valved openings along its length.

CONTINUED ▶

Closed circulatory systems

Many animals, including all vertebrates, have a closed circulatory system in which the blood is enclosed within tubes. This generates higher blood pressures as the blood is forced along fairly narrow channels instead of flowing into large cavities. This means the blood travels faster and so the blood system is more efficient at delivering substances around the body:

- The blood leaves the heart under pressure and flows along **arteries** and then **arterioles** (small arteries) to **capillaries**.

- There are extremely large numbers of capillaries. These come into close contact with most of the cells in the body, where substances are exchanged between blood and cells.

- After passing along the capillaries, the blood returns to the heart by means of **venules** (small veins) and then **veins**.

- Valves ensure that blood flows only in one direction.

Animals with closed circulatory systems are generally larger in size, and often more active than those with open systems.

Single circulatory systems

Animals with a closed circulatory system have either single circulation or double circulation. Single circulation is found, for example, in fish (Figure 1.8):

- The heart pumps deoxygenated blood to the gills.

- Here gaseous exchange takes place; there is diffusion of carbon dioxide from the blood into the water that surrounds the gills, and diffusion of oxygen from this water into the blood.

- The blood leaving the gills then flows round the rest of the body before eventually returning to the heart.

Note that the blood flows through the heart once for each complete circuit of the body.

Double circulatory systems

Birds and mammals, though, have double circulation:

- The right ventricle of the heart pumps deoxygenated blood to the lungs where it receives oxygen.

- The oxygenated blood then returns to the heart to be pumped a second time (by the left ventricle) out to the rest of the body.

This means that the blood flows through the heart twice for each complete circuit of the body. The heart gives the blood returning from the lungs an extra 'boost', which reduces the time it takes for the blood to circulate round the whole body. This allows birds and mammals to have a high metabolic rate, because oxygen and food substances required for metabolic processes can be delivered more rapidly to cells.

Q1.1 Why do only small animals have an open circulatory system?

Q1.2 What are the advantages of having a double circulatory system?

Q1.3 Fish have two-chamber hearts and mammals have four-chamber hearts. Sketch what the three-chamber heart of an amphibian, such as a frog, might look like.

Q1.4 What might be the major disadvantage of this three-chamber system?

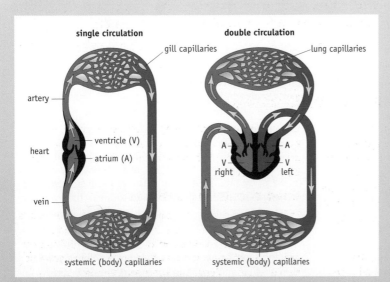

Figure 1.8 Fish have a single circulation. Birds and mammals have a double circulation.

Checkpoint ✔

1.1 Make a bullet point summary which explains why many animals have a heart and circulation.

Activity ⚙

Activities 1.3 and **1.4** let you look in detail at the structure of a mammalian heart using either a dissection or a simulation. **A1.03S** (actual dissection) **A1.04S** (simulated dissection)

How does the circulation work?

The transport medium

In the circulatory system a liquid and all the particles it contains are transported in one direction in a process known as mass flow. In animals the transport medium is usually called blood. The fluid, plasma, is mainly water and contains dissolved substances such as food, oxygen and carbon dioxide. Proteins, amino acids, salts, enzymes, hormones, antibodies and urea, the waste product from the breakdown of proteins, are just some of the substances transported in the plasma. Cells are also carried in the blood; red blood cells, white blood cells and platelets. Blood is not only important in the transport of dissolved substances and cells, but also plays a vital role in regulation of body temperature, transferring energy around the body.

Key biological principle: Properties of water that make it an ideal transport medium

Water, H_2O, is unusual among small molecules. It is a liquid at 'normal' biological temperatures; at room temperature most other small molecules, such as CO_2 and O_2, are gases. Water is a polar molecule; it has an unevenly distributed electrical charge. The two hydrogens are pushed towards each other forming a V-shaped molecule (Figure 1.9); the hydrogen end of the molecule is slightly positive and the oxygen end is slightly negative because the electrons are more concentrated at that end. It is this polarity that accounts for many of its biologically important properties.

The positively charged end of a water molecule is attracted to the negative ends of surrounding molecules. This **hydrogen bonding** holds the water molecules together and results in many of the properties of water including being liquid at room temperature.

Solvent properties

Many chemicals dissolve easily in water, allowing vital biochemical reactions to occur in the cytoplasm of cells. Free to move around in an aqueous environment, the chemicals can react, often with water itself being involved in the reactions (for example in hydrolysis and condensation reactions). The dissolved substances can also be transported around organisms, in animals via the blood and lymph systems, and in plants through the xylem and phloem.

Ionic molecules, such as sodium chloride (NaCl), dissolve easily in water. In the case of sodium chloride, the negative Cl^- ions are attracted to the positive ends of the water molecules while the positive Na^+ ions are attracted to the negative ends of the water molecules. The chloride and sodium ions become hydrated in aqueous solution, i.e. surrounded by water molecules.

Polar molecules also dissolve easily in water. Their polar groups, for example the –OH group in sugars or the amine group, $-NH_2$, in amino acids, become surrounded by water and go into solution. Such polar substances are said to be **hydrophilic** – 'water-loving'.

Non-polar, **hydrophobic** substances, such as lipids, do not dissolve in water. To enable transport in blood, lipids combine with proteins to form lipoproteins.

Thermal properties

The specific heat capacity of water, the amount of energy in joules required to raise the temperature of 1 cm³ (1 g) of water by 1 °C, is very high. This is because in water a large amount of energy is required to break the hydrogen bonds. A large input of energy causes only a small increase in temperature, so water warms up and cools down slowly. This is extremely useful for organisms, helping them to avoid rapid changes in their internal temperature and enabling them to maintain a steady temperature even when the temperature in their surroundings varies considerably.

Figure 1.9 The polarity of the water molecules results in hydrogen bonds between them.

CONTINUED ▶

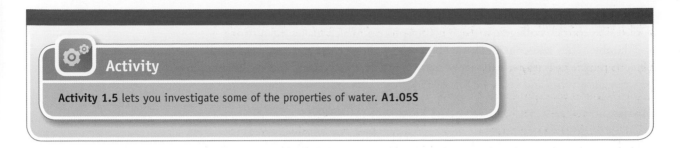

Activity

Activity 1.5 lets you investigate some of the properties of water. **A1.05S**

The heart and blood vessels

Study Figure 1.10 and locate the arteries carrying blood away from the heart and the veins returning blood to the heart.

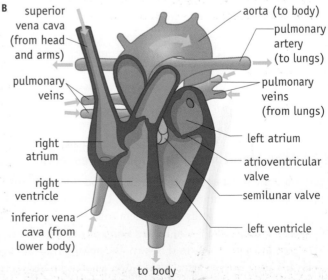

Figure 1.10 A A normal human heart. **B** Diagrammatic cross-section of the human heart (ventral or front view).

Arteries and veins can easily be distinguished, as shown in Figure 1.11. The walls of both vessels contain **collagen**, a tough fibrous protein, which makes them strong and durable. They also contain elastic fibres, which allows them to stretch and recoil. Smooth muscle cells in the walls allow them to constrict and dilate. The key differences between the arteries and veins are listed below.

Arteries:

- narrow lumen
- thicker walls
- more collagen, elastic fibres and smooth muscle
- no valves.

Veins:

- wide lumen
- thinner walls
- less collagen, elastic fibres and smooth muscle
- valves.

A

B

Figure 1.11 A Photomicrograph of an artery (left) and vein (right) surrounded by connective tissue. **B** Diagram of an artery, a vein and a capillary. The endothelium that lines the blood vessels is made up of epithelial cells (see page 57).

The capillaries that join the small arteries (arterioles) and small veins (venules) are very narrow, about 10 μm in diameter, with walls only one cell thick.

These features can be directly related to the functions of the blood vessels, as described below.

Activity

Activity 1.6 lets you investigate how the structure of blood vessels relates to their function. **A1.06S**

How does blood move through the vessels?

Every time the heart contracts (**systole**), blood is forced into arteries and their elastic walls stretch to accommodate the blood. During **diastole** (relaxation of the heart), the elasticity of the artery walls causes them to recoil behind the blood, helping to push the blood forward. The blood moves along the length of the artery as each section in series stretches and recoils in this way. The pulsing flow of blood through the arteries can be felt anywhere an artery passes over a bone close to the skin.

Checkpoint

1.2 Identify the key structures of an artery, a vein and a capillary, and in each case explain how the structure is related to the function of the vessel.

By the time the blood reaches the smaller arteries and capillaries there is a steady flow of blood. In the capillaries this allows exchange between the blood and the surrounding cells through the one-cell-thick capillary walls. The network of capillaries that lies close to every cell ensures that there is rapid diffusion between the blood and surrounding cells.

The heart has a less direct effect on the flow of blood through the veins. In the veins blood flow is assisted by the contraction of skeletal muscles during movement of limbs and breathing. Low pressure developed in the thorax (chest cavity) when breathing in also helps draw blood back into the heart from the veins. Backflow is prevented by valves within the veins (Figure 1.12). The steady flow without pulses of blood means that the blood is under low pressure in veins.

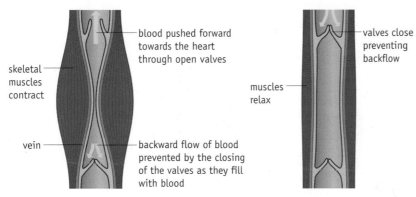

Figure 1.12 Valves in the veins prevent the backflow of blood.

Q1.5   List the features shown in Figure 1.11B that enable the artery to withstand high pressure and then recoil to maintain a steady flow of blood.

Since the heart is a muscle it needs a constant supply of fresh blood. You might think that receiving a blood supply would never be a problem for the heart. However, the heart is unable to use any of the blood inside its pumping chambers directly. Instead, the heart muscle is supplied with blood through two vessels called the **coronary arteries**. You can see the coronary arteries and coronary veins on the surface of the heart in Figure 1.10A.

How the heart works

Give a tennis ball a good, hard squeeze. You're using about the same amount of force that your heart uses in a single contraction to pump blood out to the body. Even when you are at rest, the muscles of your heart work hard – weight for weight, harder than the leg muscles of a person running.

The chambers of the heart alternately contract (systole) and relax (diastole) in a rhythmic cycle. One complete sequence of filling and pumping blood is called a **cardiac cycle**, or heartbeat. During systole, cardiac muscle contracts and the heart pumps blood out through the aorta and pulmonary arteries. During diastole, cardiac muscle relaxes and the heart fills with blood.

The cardiac cycle can be simplified into three phases: atrial systole, ventricular systole and diastole. The events that occur during each of the stages are shown in Figure 1.13.

Phase 1: Atrial systole

Blood returns to the heart due to the action of skeletal and gaseous exchange (breathing) muscles as you move and breathe. Blood under low pressure flows into the **left** and **right atria** from the pulmonary veins and vena cava. As the atria fill, the pressure of blood against the **atrioventricular valves** pushes them open and blood begins to leak into the **ventricles**. The atria walls contract, forcing more blood into the ventricles. This is known as atrial systole.

Phase 2: Ventricular systole

Atrial systole is immediately followed by ventricular systole. The ventricles contract from the base of the heart upwards, increasing the pressure in the ventricles. This pushes blood up and out through the arteries. The pressure of blood against the atrioventricular valves closes them and prevents blood flowing backwards into the atria.

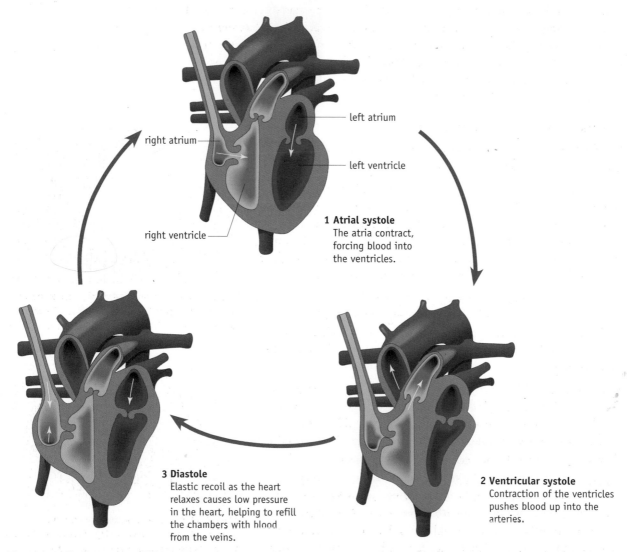

right atrium

left atrium

left ventricle

right ventricle

1 Atrial systole
The atria contract, forcing blood into the ventricles.

3 Diastole
Elastic recoil as the heart relaxes causes low pressure in the heart, helping to refill the chambers with blood from the veins.

2 Ventricular systole
Contraction of the ventricles pushes blood up into the arteries.

Figure 1.13 The three stages of the cardiac cycle. At each stage blood moves from higher to lower pressure.

Phase 3: Diastole

The atria and ventricles then relax during diastole. Elastic recoil of the relaxing heart walls lowers pressure in the atria and ventricles. Blood under higher pressure in the arteries is drawn back towards the ventricles, closing the **semilunar valves** and preventing further backflow. The coronary arteries fill during diastole. Low pressure in the atria helps draw blood into the heart from the veins.

Closing of the atrioventricular valves and then the semilunar valves creates the characteristic sounds of the heart.

Q1.6 When the heart relaxes in diastole, you might expect blood to move from the arteries back into the ventricles due to the elastic recoil of the heart and the action of gravity if you are standing or sitting upright. How is this prevented?

Checkpoint ✔

1.3 Make a flowchart which summarises the events in the cardiac cycle.

Activity ⚙

Activity 1.7 lets you test your knowledge of the cardiac cycle.
A1.07S

What is atherosclerosis?

Atherosclerosis is the disease process that leads to coronary heart disease and strokes. In atherosclerosis fatty deposits can either block an artery directly, or increase its chance of being blocked by a blood clot (**thrombosis**). The blood supply can be blocked completely. If this happens for long, the affected cells are permanently damaged. In the arteries supplying the heart this results in a heart attack (**myocardial infarction**); in the arteries supplying the brain it results in a stroke. The supply of blood to the brain is restricted or blocked, causing damage or death to cells in the brain. Narrowing of arteries to the legs can result in tissue death and gangrene (decay). An artery can burst where blood builds up behind an artery narrowed as a result of atherosclerosis.

What happens in atherosclerosis?

Atherosclerosis can be triggered by a number of factors. Whatever the trigger, this is the course of events that follows:

1 The endothelium, a delicate layer of cells that lines the inside of an artery (Figure 1.14A), separating the blood that flows along the artery from the muscular wall, becomes damaged for some reason. For instance, this endothelial damage can result from high blood pressure, which puts an extra strain on the layer of cells, or it might result from some of the toxins from cigarette smoke in the bloodstream.

2 Once the inner lining of the artery is breached, there is an inflammatory response. White blood cells leave the blood vessel and move into the artery wall. These cells accumulate chemicals from the blood, particularly **cholesterol**. A deposit builds up, called an **atheroma**.

3 Calcium salts and fibrous tissue also build up at the site, resulting in a hard swelling called a **plaque** on the inner wall of the artery. The build-up of fibrous tissue means that the artery wall loses some of its elasticity; in other words, it hardens. The ancient Greek word for 'hardening' is 'sclerosis', giving the word 'atherosclerosis'.

4 Plaques cause the artery to become narrower (Figure 1.14B). This makes it more difficult for the heart to pump blood around the body and can lead to a rise in blood pressure. Now there is a dangerous **positive feedback** building up. Plaques lead to raised blood pressure and raised blood pressure makes it more likely that further plaques will form.

Figure 1.14 A Photomicrograph of a normal, healthy coronary artery showing no thickening of the arterial wall. The lumen is large. Magnification ×215.

Figure 1.14 B Photomicrograph of a diseased coronary artery showing narrowing of the lumen (blue) due to atheroma deposits (pink cells) and build-up of atherosclerotic plaque (yellow). Magnification ×230.

The person is probably unaware of any problem at this stage, but if the arteries become very narrow or completely blocked then they cannot supply enough blood to bring oxygen and nutrients to the tissues. The tissues can no longer function normally and symptoms will soon start to show.

Why does the blood clot in arteries?

When blood vessel walls are damaged or blood flows very slowly, a blood clot is much more likely to form (Figure 1.15). When **platelets**, a type of blood cell without a nucleus, come into contact with the damaged vessel wall they change from flattened discs to spheres with long thin projections (Figure 1.16). Their cell surfaces change, causing them to stick to the exposed collagen in the wall and to each other to form a temporary platelet plug. They also release substances that activate more platelets.

Figure 1.15 Photomicrograph of a diseased coronary artery showing narrowing and a blood clot. Magnification ×245.

Figure 1.16 Electron micrograph showing activated platelets. Magnification ×6000.

The direct contact of blood with collagen within the damaged artery wall also triggers a complex series of chemical changes in the blood (Figure 1.17). A **cascade** of changes results in the soluble plasma protein called **prothrombin** being converted into **thrombin**. Thrombin is an **enzyme** that catalyses the conversion of another soluble plasma protein, **fibrinogen**, into long insoluble strands of the protein **fibrin**. These fibrin strands form a tangled mesh that traps blood cells to form a clot (Figures 1.17 and 1.18).

Activity

Activity 1.8 lets you summarise the steps in development of atherosclerosis and clot formation. **A1.08S**

Why do only arteries get atherosclerosis?

The fast-flowing blood in arteries is under high pressure so there is a significant chance of damage to the walls. The low pressure in the veins means that there is less risk of damage to the walls.

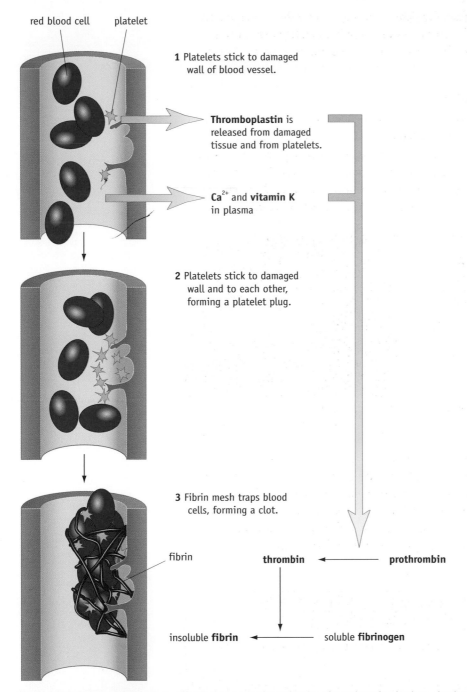

red blood cell   platelet

1 Platelets stick to damaged wall of blood vessel.

Thromboplastin is released from damaged tissue and from platelets.

Ca^{2+} and **vitamin K** in plasma

2 Platelets stick to damaged wall and to each other, forming a platelet plug.

3 Fibrin mesh traps blood cells, forming a clot.

fibrin

thrombin ◄——————— **prothrombin**

insoluble **fibrin** ◄——————— soluble **fibrinogen**

Figure 1.17 Damage to the vessel walls triggers a complicated series of reactions that leads to clotting.

Figure 1.18 False-colour scanning electron micrograph showing red blood cells and platelets (green) trapped in the yellow mesh of fibrin. Magnification ×1850.

The consequences of atherosclerosis

Coronary heart disease

Narrowing of the coronary arteries limits the amount of oxygen-rich blood reaching the heart muscle. The result may be a chest pain called **angina**. Angina is usually experienced during exertion. Because the heart muscle lacks oxygen, it is forced to respire **anaerobically**. It is thought that this results in chemical changes which trigger pain but the detailed mechanism is still not known.

If a fatty plaque in the coronary arteries ruptures, cholesterol is released which leads to rapid clot formation. The blood supply to the heart may be blocked completely. The heart muscle supplied by these arteries does not receive any blood, so it is said to be ischaemic (without blood). If the affected muscle cells are starved of oxygen for long they will be permanently damaged. This is what we call a heart attack or myocardial infarction. If the zone of dead cells occupies only a small area of tissue, the heart attack is less likely to prove fatal.

Stroke

If the supply of blood to the brain is only briefly interrupted then a mini-stroke may occur. A mini-stroke has all the symptoms of a full stroke but the effects last for only a short period, and full recovery can happen quite quickly. However, a mini-stroke is a warning of problems with blood supply to the brain that could result in a full stroke in the future.

Did you know?

The symptoms of cardiovascular disease

Coronary heart disease

Shortness of breath and angina are often the first signs of coronary heart disease. The symptoms of angina are intense pain, an ache or a feeling of constriction and discomfort in the chest or in the left arm and shoulder. Other symptoms are unfortunately very similar to those of severe indigestion and include a feeling of heaviness, tightness, pain, burning and pressure – usually behind the breastbone, but sometimes in the jaw, arm or neck. Women may not have chest pain but experience unusual fatigue, shortness of breath and indigestion-like symptoms.

Sometimes coronary heart disease causes the heart to beat irregularly. This is known as **arrhythmia** and can itself lead to heart failure. Arrhythmia can be important in the diagnosis of coronary heart disease.

Stroke

The effects of a stroke will vary depending on the type of stroke, where in the brain the problem has occurred and the extent of the damage. The more extensive the damage, the more severe the stroke and the lower the chance of full recovery. The symptoms normally appear very suddenly and include:

- numbness
- dizziness
- confusion
- slurred speech
- blurred or lost vision, often only in one eye.

Visible signs often include paralysis on one side of the body with a drooping arm, leg or eyelid, or a dribbling mouth. The *right* side of the brain controls the *left* side of the body, and vice versa; therefore the paralysis occurs on the opposite side of the body to where the stroke occurred.

? Did you know?

Aneurysms

If part of an artery has narrowed and become less flexible, blood can build up behind it. The artery bulges as it fills with blood and an **aneurysm** forms. An atherosclerotic aneurysm of the aorta is shown in Figure 1.19.

What will eventually happen as the bulge enlarges and the walls of the aorta are stretched thin? Aortic aneurysms are likely to rupture when they reach about 6–7 cm in diameter. The resulting blood loss and shock can be fatal. Fortunately, earlier signs of pain may prompt a visit to the doctor. The bulge can often be felt in a physical examination or seen with ultrasound examination and it may be possible to surgically replace the damaged artery with a section of artificial artery.

Figure 1.19 An aneurysm in the aorta below the kidneys. If an aneurysm ruptures it can be fatal.

Extension

Read **Extension 1.1** to find out how you may be able to save someone's life by carrying out cardiopulmonary resuscitation. **X1.01S**

Extension

There are several tests used to diagnose cardiovascular disease that can be requested by doctors and you can read more details of these tests in **Extension 1.2**. **X1.02S**

1.2 Who is at risk of cardiovascular disease?

Probability and risk

What do we mean by risk?

Risk is defined as 'the probability of occurrence of some unwanted event or outcome'. It is usually in the context of hazards, that is, anything that can potentially cause harm, such as the chance of contracting lung cancer if you smoke. Probability has a precise mathematical meaning and can be calculated to give a numerical value for the size of the risk. Do not panic – the maths is simple!

Taking a risk is a bit like throwing a die (singular of 'dice'). You can calculate the chance that you will have an accident or succumb to a disease (or throw a six). You will not *necessarily* suffer the accident or illness, but by looking at past circumstances of people who have taken the same risk, you can estimate the chance that you will suffer the same fate to a reasonable degree of accuracy.

Working out probabilities

There are six faces on a standard die. Only one face has six dots, so the chance of throwing a six is 1 in 6 (provided the die is not loaded). Scientists tend to express '1 in 6' as a decimal: 0.166666 recurring (about 0.17). In other words, each time you throw a standard die, you have about a 0.17 or 17% chance of throwing a one, about a 17% chance of throwing a two, and so on.

When measuring risk you must always quote a time period for the risk. Here you have a 17% chance of throwing a one with each throw of the die.

In a Year 5 class of 30 pupils, six children caught head lice in one year. The risk of catching head lice in this class was therefore 6 in 30 or 1 in 5, giving a probability of 0.2 or 20% in a year.

Estimating risks to health

In 2005, 19 429 people in the UK died due to injuries or poisoning. The total UK population at the time was 60 209 408, so we can calculate the average risk in a year of someone in the UK dying from injuries or poisoning as:

19 429 in 60 209 408

or 1 in $\dfrac{60\,209\,408}{19\,429}$

= 1 in 3099

= $\dfrac{1}{3099}$

= 0.000 32 or 0.032%

Another way of working this out is as:

$\dfrac{19\,429}{60\,209\,408}$

However, when calculating a probability in relation to health, most people would find, for example, 1 in 3099 more meaningful than 0.000 32 or 0.032%.

Assuming the proportion of people that die from injuries or poisoning remains much the same each year, this calculation gives an estimate of the risk for any year.

If we calculated the risk of any one of us developing lung cancer in our lifetime we would find a probability of 1 in about 1600. However, because lung cancer is much more likely if you smoke, the risk for smokers is far greater. When looking at calculated risk values you need to think about exposure to the hazard.

Q1.7 Look at the causes of death listed below and put them in order, from the most likely to the least likely. You could also have a go at estimating the probability of someone in the UK dying from each cause during a year.

- accidental poisoning
- heart disease
- injury purposely inflicted by another person
- lightning
- lung cancer
- railway accidents
- road accidents

Did you get it right?

People frequently get it wrong, underestimating or overestimating risk. We can say that there is about a 1 in 1700 risk of each of us dying from lung cancer in any one year, a 1 in 100 000 risk of our being murdered in the next 12 months, and a 1 in 10 million risk of our being hit by lightning in a year. However, recent work on risk has concentrated not so much on numbers such as these but on the perception of risk.

Activity

Activity 1.9 asks you to estimate risks for a range of diseases using National Office for Statistics data. **A1.09S**

Perception of risk

The significance of the perception of risk can be illustrated by a decision in September 2001 made by the American Red Cross, which provides about half of the USA's blood supplies. They decided to ban all blood donations from anyone who has spent six months or more in any European country since 1980. Their reason was the risk of transmitting variant Creutzfeldt–Jakob disease (vCJD), the human form of bovine spongiform encephalopathy (BSE), through blood transfusion. Experts agreed that there was a *chance* of this happening. Yet there wasn't a single known case of its *actually* having happened. Indeed, as the USA is short of blood for blood transfusions, it is possible that more people may have died as a result of this 'safety precaution' than would have been the case without it.

So why did America ban European blood donations? The likely reason was public perceptions of the risk of contracting vCJD. People will *overestimate* the risk of something happening if the risk is:

- involuntary (not under their control)
- not natural
- unfamiliar
- dreaded
- unfair
- very small.

If you look at this list you should be able to see why people may greatly overestimate some risks (such as the chances of contracting vCJD from blood transfusions) while underestimating others (such as the dangers of driving slightly faster than the speed limit or playing on a frozen lake).

Nowadays many risk experts argue that perceptions of risk are what really drive people's behaviour. Consider what happened when it became compulsory in the UK to use seat belts for children in the rear seats of cars (Figure 1.20). The number of children killed and injured *increased*. How could this be? John Adams, an academic at University College London, argues that this is because the parents driving felt safer once their children were wearing seat belts and so drove slightly less carefully. Unfortunately, this change in their driving behaviour was more than enough to compensate for any extra protection provided by the seat belts.

There is a tendency to overestimate the risks of sudden imposed dangers where the consequences are severe, and to *underestimate* a risk if it has an effect in the long-term future, even if that effect is severe, for example, the health risks associated with smoking or poor diet.

A useful distinction is sometimes made between risk and uncertainty. When we lack the data to estimate a risk precisely, we are *uncertain* about the risk. For example, we are uncertain about the environmental consequences of many chemicals.

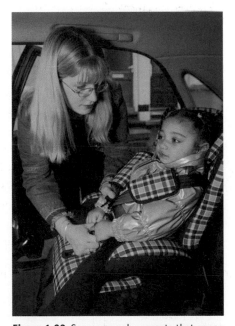

Figure 1.20 Some research suggests that young children who wear rear seat belts are more likely to die in an accident than those who don't. But this may be explained by parents' driving habits. Health risks are greatly affected by human behaviour.

Q1.8 In a school of 1300 students, in one term 10 students contracted verrucas from the school pool. In a letter to parents the head teacher said there was a less than 1% chance of any child catching a verruca in any term. Was the figure she quoted correct and what assumptions had she made in making this statement?

Q1.9 In 2005, 109 958 cases of chlamydia were reported in the UK, with 21 215 of these cases being reported in London. One paper wanted to write a front page headline claiming that there was a higher risk of contracting this sexually transmitted infection in the capital compared to the rest of the country. Would they have been correct? Support your answer with calculated risk values. The population of the UK in 2005 was 60.2 million; the population of London was 7.5 million.

Different types of risk factor

In the UK the estimated risk of any one of us having fatal heart disease in any one year is about 1 in 600 compared to 1 in 1050 for a fatal stroke. However, these probabilities use figures for the whole population, giving averages which make the simplistic assumption that everyone has the same chance of having cardiovascular disease. This is obviously not the case.

The averages take no account of any risk factors – things that increase the chance of the harmful outcome. When assessing an individual's risk of bad health, all the contributing risk factors need to be established.

There are many different factors that contribute to health risks, for example:

- heredity
- physical environment
- social environment
- lifestyle and behaviour choices.

Identifying risk factors – correlation and causation

To determine what the risk factors are for a particular disease, scientists look for **correlations** between potential risk factors and the occurrence of the disease.

Two variables are positively correlated when an increase in one is accompanied by an increase in the other (Figure 1.21A). For example, the length of a TV programme and the percentage of the class asleep might be positively correlated. The number of cigarettes smoked over a lifetime and the chance of developing cardiovascular disease certainly are. If the values of one variable decrease while the other increases, there is a negative correlation (Figure 1.21B).

Large amounts of data are needed to ensure that the correlation is statistically significant; in other words, not just an apparent correlation due to chance.

It is important to realise that a correlation between two variables does not necessarily mean that the variables are causally linked. Two variables are causally linked when a change in one is responsible for a change in the other. It is easy to think of variables that are correlated where there is no **causation**. For example, worldwide, speaking English as your first language correlates quite well with having a greater-than-average life expectancy. This, though, is simply because countries like the USA, UK, Australia and Canada have a higher-than-average standard of living. It is this that causes increased life expectancy through better nutrition, medical care and so on, rather than the language spoken.

It is because of this logical gap between correlation and causation that scientists try, whenever they can, to carry out experiments in which they can control variables, to see if altering one variable really does have the predicted effect. To do this, scientists often set up a **null hypothesis**. They assume for the sake of argument that there will be no difference between an experimental group and a control group, and then test this hypothesis using statistical analysis.

A Positive correlation

Variable 1 / Variable 2

B Negative correlation

Variable 1 / Variable 2

Figure 1.21 A When an increase in one variable is accompanied by an increase in the other, there is a positive correlation, giving a scattergram rising from left to right. **B** With a negative correlation, one set of data increases while the other falls, resulting in a graph going down from left to right.

Q1.10 Strong correlations have been reported between the following pairs of variables. In each case, decide if there is likely to be a causal link between the variables or not. Suggest a possible reason for the correlation.

a shark attacks and ice cream sales
b children's foot sizes and their spelling abilities
c lung cancer and smoking.

1.3 Risk factors for cardiovascular disease

Identifying risk factors for CVD

Large-scale studies have been undertaken to find the risk factors for many common diseases, including cardiovascular disease. Epidemiologists, scientists who study patterns in the occurrence of disease, look for correlations between a disease and specific risk factors.

Two commonly used designs for this type of study are:

- cohort studies – a group of people are followed over time to see who develops the disease
- case-control studies – a group of people who have the disease are compared with a group who do not have the disease.

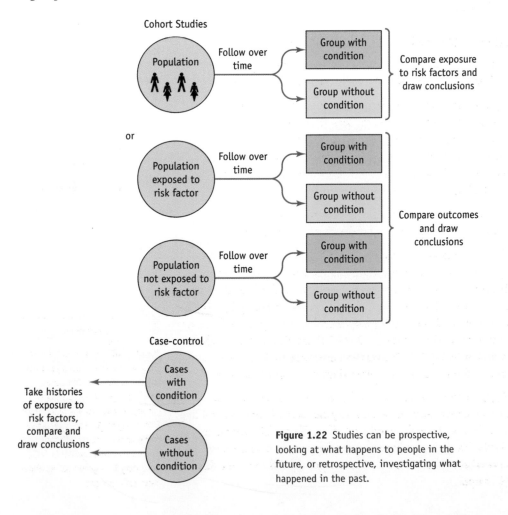

Figure 1.22 Studies can be prospective, looking at what happens to people in the future, or retrospective, investigating what happened in the past.

Cohort studies

Cohort studies follow a group of people over time to see who develops the disease and who does not. During the study people's exposure to suspected risk factors is recorded so any correlations between the risk factors and disease development can be identified. It may take a long time for the condition to develop so these studies can take years and be very expensive.

The first major cohort study into CVD started in 1948. At the time, little was known about the causes of heart disease and stroke. The aim of the Framingham Heart Study was to identify the factors that contribute to the development of the disease. A random sample of 5209 men and women between the ages of 30 and 62 from the town of Framingham, Massachusetts, was recruited for the study. At the time of recruitment they had no symptoms of cardiovascular disease. In 1971, 5124 of the participants' adult children joined the study. 3900 of their grandchildren joined the study in 2002.

Every two years the participants are asked to provide a detailed medical history, undergo a physical examination and tests, and answer questions about their lifestyle. The data are used to look for common features that contribute to the development of CVD. High blood pressure, high blood cholesterol, smoking, obesity, diabetes and physical inactivity were all identified as major CVD risk factors as a result of this study.

Other studies have confirmed these findings. The World Health Organization MONICA study (MONItoring trends and determinants in CArdiovascular disease), involving over 7 million people in 21 countries over 10 years, confirmed the link between several of these factors and increased occurrence of the disease.

Case-control studies

In a case-control study a group of people with a disease (cases) are compared with a control group of individuals who do not have the disease. Information is collected about the risk factors that they have been exposed to, allowing factors that may have contributed to development of the disease to be identified.

The control group should be representative of the population from which the case group was drawn. Sometimes controls are individually matched to cases; known disease-risk factors, such as age and sex, are then similar in each case and control pair. This allows scientists to investigate the potential role of unknown risk factors. It should be noted that factors used to match the cases and controls cannot be investigated within the study, so it is important not to match on variables which could potentially turn out to be risk factors.

One of the first case-control studies was conducted in the 1950s by two British scientists, Richard Doll and Austin Bradford Hill, to determine if there was a link between smoking and lung cancer. A group of hospital patients with lung cancer was compared with a second group who did not have cancer. The data indicated a correlation between smoking and lung cancer.

Q1.11 Did the study by Doll and Hill prove that smoking caused lung cancer? Give a reason for your answer.

Activity

In **Activity 1.10** you evaluate the design of studies used to determine health risk factors. **A1.10S**

Features of a good study

To identify correlations between risk factors and disease, studies need to be carefully designed. Recording a higher rate of heart disease in 50 people who drink more alcohol than the recommended amount, compared with 50 people who drink less than the recommended amount, supports the suggestion that excess alcohol consumption increases the risk of developing heart disease. However, the group who drink more alcohol might also smoke more, do less exercise and eat a fatty diet. Any of these factors could be linked to developing the disease. A well-designed study tries to overcome these problems. When designing an epidemiological study, there are some key questions that should be considered.

Clear aim

A well-designed study should include a clearly stated hypothesis or aim. The design of the study must be appropriate to the stated hypothesis or aim and produce results that are valid and reliable.

Representative sample

A representative sample must be selected from the wider population that the study's conclusions will be applied to. Selection bias occurs when those who participate in a study are not representative of the target population. For example, if a study aiming to look at the prevalence of disease in a community only sent out questionnaires to people on the voting register, the findings may not be representative. This is because people under 18 years old, people who had recently moved in or out of the area, and people in temporary accommodation would be missed.

Differences between people asked to take part in a study and those who actually respond should also be considered before generalising findings to the target population. Non-participants can differ in important respects from participants. For example, if a study involves interviewing people at home during the day, then those employed outside the home may be less likely to participate. The health and lifestyle of employed and non-employed people differs in many ways, so the findings could be misleading.

The proportion of individuals who drop out of a study after it has begun should be kept to a minimum. This is particularly important in cohort studies which follow people over long periods of time. People who drop out of studies often share common features. It is important to monitor the characteristics of the remaining participants to ensure that they are still representative of the target population.

Valid and reliable results

Any methods used must produce valid data, from measurements that provide information on what the study set out to measure. If studying the effect of blood pressure on development of CVD, valid blood pressure measurements would be made using an appropriate blood pressure monitor. A survey to study the effect of alcohol consumption on the development of coronary heart disease could introduce problems with validity if it relied on the participants recalling the quantity of alcohol they consumed. Participants may not recall correctly because they were intoxicated, or they may underestimate because they are reluctant to admit their true consumption.

The method used to collect results must be reliable. A reliable method used at different times, or by different people, will produce similar results. A reliable test will also give similar results for repeated measurements. If measuring blood pressure, the same type of equipment and same procedure should be used each time the measurement is made. Any variables that could affect the measurement should be controlled or taken into account. A method using questionnaires, for example to conduct a survey on lifestyle factors, should use the same questions for each participant.

The disease diagnosis must be clearly defined, to ensure that different doctors record and measure symptoms in the same way. The development of coronary heart disease or onset of Alzheimer's, for example, must be measured and recorded using standard methods which are the same for all participants in the study.

A sample must be large enough to produce results that could not have occurred by chance. In cohort studies of a rare disease, only a small proportion of the population will develop the disease. In case-control studies, only a few people may have been exposed to the factors under investigation, or, in the case of rare diseases, the number of cases may be low to start with. With larger samples, more reliable estimates for the wider population can be calculated. For a condition that affects 5% of the population each year, a cohort of 1000 people would need to be followed for 10 years in order to collect information on 50 people with the disease. Similarly, in case-control studies, sufficient participants need to be recruited in order to detect any effects due to rare exposures.

The potential effect of all variables that could be correlated with the disease should be considered when designing the study. For example, in a study of blood pressure and development of CVD, a group of people with low blood pressure is compared to a group with higher blood pressure. The data shows that the group with lower blood pressure has less CVD. However, if the average age of this group is less than the high pressure group, the difference in CVD development may be due to the age difference and not blood pressure. Age is a factor known to be associated with CVD. Matching case and control groups on variables known to correlate with the disease being studied will ensure that only the factor under investigation is influencing the outcome.

> **Checkpoint** ✔
>
> **1.5** Produce a checklist of the features of a well-designed health risk study that ensure valid and reliable data is collected.

Q1.12 The Framingham cohort is primarily white. Explain whether it would be valid to extrapolate these results to the general population of the USA.

Risk factors for CVD

Your chances of having coronary heart disease or a stroke are increased by several inter-related risk factors, the majority of which are common to both conditions. These include:

- high blood pressure
- obesity
- blood cholesterol and other dietary factors
- smoking
- genetic inheritance.

Some of these you can control, while others you can't.

Figure 1.23 Some of the potential risk factors for developing coronary heart disease are easy to identify, but may be difficult to control.

Age and gender make a difference

Q1.13 Look at Table 1.1. What happens to your risk of developing cardiovascular disease as you get older?

Q1.14 Does this mean that, at your age, you need not worry?

Age/years	Female		Male	
	Heart attack	Stroke	Heart attack	Stroke
16–44	0	0	0	1
45–64	18	5	18	8
65–74	51	12	84	17
75 and over	67	28	83	54

Table 1.1 Rates per 1000 population reporting longstanding diseases of the circulatory system by sex and age, 2004, Great Britain. *Source: Office for National Statistics General Household Survey.*

Q1.15 Look at Table 1.2. Do these data suggest that males and females face the same risk of cardiovascular disease?

Age/years	Female deaths	Male deaths
under 1	13	22
1–4	8	12
5–14	19	29
15–24	65	96
25–34	170	341
35–44	552	1343
45–54	1362	3930
55–64	3541	9330
65–74	11 304	19 783
75–84	35 958	35 607
85+	46 470	20 816
Totals	99 462	91 309

Table 1.2 Mortality data from diseases of the circulatory system for England and Wales in 2004. *Source: Office for National Statistics Deaths by age, sex and underlying cause, 2004 registrations.*

Q1.16 Comment on the reliability of the data presented in Tables 1.1 and 1.2.

Q1.17 Many people now think that, until menopause, a woman's reproductive hormones offer her protection from coronary heart disease. Do these data support this view? Is it valid to draw this conclusion from these data?

Activity

In **Activity 1.11** you compare data for coronary heart disease and stroke and look at trends over a ten-year period. **A1.11S**

The risk of cardiovascular disease is higher for men than women in the UK. For a man aged 55, the risk of a heart attack before he is 60 is about 2%, 1 in 50, whereas a woman aged 55 has a risk of 1 in 100 of having a heart attack by the time she is 60. In both sexes, the prevalence of cardiovascular disease increases with age. This may be due to the effects of ageing on the arteries; they tend to become less elastic and may be more easily damaged. With increasing age the risks associated with other factors may increase, causing a rise in the number of cases of disease.

High blood pressure

Elevated blood pressure, known as **hypertension**, is considered to be one of the most common factors in the development of cardiovascular disease. High blood pressure increases the likelihood of atherosclerosis occurring.

Blood pressure is a measure of the hydrostatic force of the blood against the walls of a blood vessel. You should remember that blood pressure is higher in arteries and capillaries than in veins. The pressure in an artery is highest during the phase of the cardiac cycle when the ventricles have contracted and forced blood into the arteries. This is the **systolic pressure**. Pressure is at its lowest in the artery when the ventricles are relaxed. This is the **diastolic pressure**.

Measuring blood pressure

A **sphygmomanometer** is a traditional device used to measure blood pressure. It consists of an inflatable cuff that is wrapped around the upper arm, and a manometer or gauge that measures pressure (Figure 1.24). When the cuff is inflated the blood flow through the artery in the upper arm is stopped. As the pressure in the cuff is released the blood starts to flow through the artery. This flow of blood can be heard using a stethoscope positioned on the artery below the cuff. A pressure reading is taken when the blood first starts to spurt through the artery that has been closed. This is the *systolic* pressure. A second reading is taken when the pressure falls to the point where no sound can be heard in the artery. This is the *diastolic* pressure.

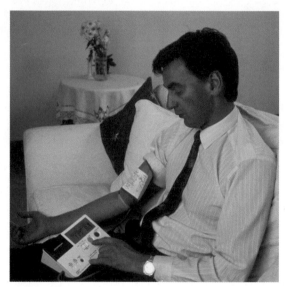

Figure 1.24 Nowadays blood pressure monitors can give digital readouts.

The SI units (International System of Units) for pressure are kilopascals, but in medical practice it is traditional to use millimetres of mercury, mmHg. (The numbers refer to the number of millimetres the pressure will raise a column of mercury.)

Blood pressure is reported as two numbers, one 'over' the other, for example $\frac{140}{85}$ (140 over 85). This means a systolic pressure of 140 mmHg and a diastolic pressure of 85 mmHg. For an average, healthy person you would expect a systolic pressure of between 100 and 140 mmHg and a diastolic pressure of between 60 and 90 mmHg.

systolic pressure, the maximum blood pressure when the heart contracts	→ $\dfrac{140}{85}$ ←	diastolic pressure, the blood pressure when the heart is relaxed

Peter's blood pressure was an incredible $\frac{240}{140}$.

Activity

In **Activity 1.12** you use a sphygmomanometer, a blood pressure monitor or the accompanying simulation to measure blood pressure. **A1.12S**

What determines your blood pressure?

Contact between blood and the walls of the blood vessels causes friction, and this impedes the flow of blood. This is called peripheral resistance. The arterioles and capillaries offer a greater total surface area, resisting flow more, slowing the blood down and causing the blood pressure to fall. Notice in Figure 1.25 that the greatest drop in pressure occurs in the arterioles. The fluctuations in pressure in the arteries are caused by contraction and relaxation of the heart. As blood is expelled from the heart, pressure is higher. During diastole, elastic recoil of the blood vessels maintains the pressure and keeps the blood flowing.

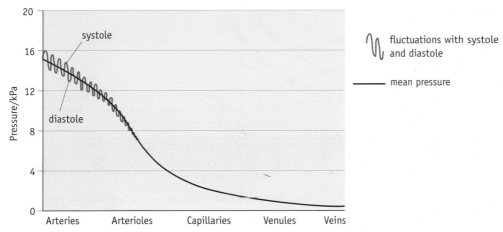

Figure 1.25 Blood pressure in the circulatory system. As peripheral resistance increases with greater total surface area, the flow of blood slows causing pressure to fall.

Activity

Draw a **concept map** for blood pressure to bring together all the ideas covered. A pro-forma is available in **Activity 1.13** if you don't want to start from scratch. **A1.13S**

If the smooth muscles in the walls of an artery or an arteriole contract, the vessels constrict, increasing resistance. In turn, your blood pressure is raised. If the smooth muscles relax, the lumen is dilated, so peripheral resistance is reduced and blood pressure falls. Any factor that causes arteries or arterioles to constrict can lead to elevated blood pressure. Such factors include natural loss of elasticity with age, release of hormones such as adrenaline, or a high-salt diet. In turn high blood pressure can lead to atherosclerosis.

One sign of high blood pressure is **oedema**, fluid building up in tissues and causing swelling. Oedema may also be associated with kidney or liver disease, or with restricted body movement.

At the arterial end of a capillary, blood is under pressure. This forces fluid and small molecules normally found in plasma out through the capillary walls into the intercellular spaces, forming **tissue fluid** (also called interstitial fluid) (Figure 1.26). The capillary walls prevent blood cells and larger plasma proteins from passing through, so these stay inside the capillaries.

If blood pressure rises above normal, more fluid may be forced out of the capillaries. In such circumstances, fluid accumulates within the tissues causing oedema.

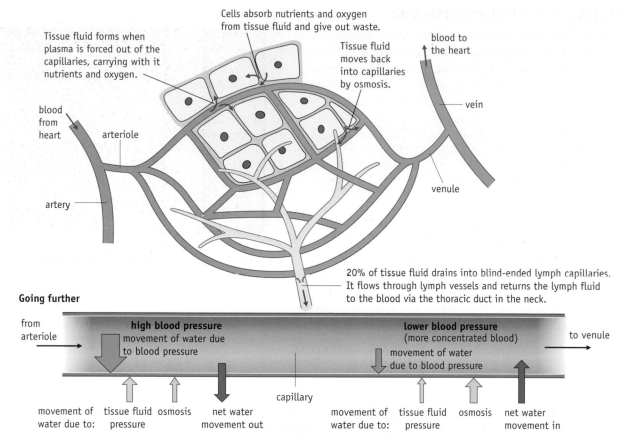

Figure 1.26 Production of tissue fluid in a capillary bed. For details on osmosis see page 72.

Q1.18 During left-side heart failure (the most frequent type) there is an increase in pressure in the pulmonary vein and left atrium. This is because blood continues to flow out of the right side of the heart to the lungs and return to the heart due to the action of breathing muscles. Where in the body will blood pressure rise and oedema form?

Dietary factors

Our choices of food, in particular the type and quantity of high-energy food, can either increase or decrease our risk of developing certain diseases, including cardiovascular diseases.

Carbohydrates, **lipids** (often called fats and oils) and **proteins** are constituents of our food which store energy. Alcohol can also provide energy. The relative energy content of these nutrients is shown in Table 1.3.

Nutrient	Energy available per gram/kJ
Carbohydrates	16
Lipids	37
Proteins	17
Alcohol	29

Table 1.3 Energy content of nutrients

Energy units – avoiding confusion

Most packet foods these days detail the energy content per 100 g or other appropriate quantity. In Figure 1.27, notice the units used to express energy content for a bar of chocolate. Why are two different units quoted? Which should we use?

Traditionally, energy was measured in **calories**; one calorie is the quantity of heat energy required to raise the temperature of 1 cm^3 of water by 1 °C. Food labels normally display units of 1000 calories, called **kilocalories** (also called **Calories** with a capital C).

The SI unit for energy is the joule (J), and 4.18 joules = 1 calorie. The **kilojoule** (1 kJ = 1000 joules) is used extensively in stating the energy contents of foods. In the popular press the Calorie is still used as the basic unit of energy, particularly with reference to weight control. Hence most food labels in the UK continue to quote both Calories and kilojoules (Figure 1.27).

Nutrition Information		Per Bar	Per 100g
Energy	kJ	880	2260
	kcal	210	540
Protein	g	2.7	7.0
Carbohydrate	g	20.7	53.8
Fat	g	12.8	33.2

Ingredients: milk, sugar, cocoa butter, cocoa

Figure 1.27 How much energy does this chocolate contain? Notice that the label displays energy values in kilojoules and kilocalories.

Q1.19 A newborn baby requires around 2000 kJ per day. Express this **a** in calories **b** in Calories.

Did you know?

The role of culture in our eating habits

When a baby is born it survives entirely on milk (from the breast or a bottle) for a period that lasts from a couple of months to a year or more. Mother's milk is pretty much the same the world over. Aside from being a bit low in iron, fibre and some vitamins (e.g. vitamin C) it's a near perfect human diet.

Once weaning starts, a baby starts to take in solid food as well. This is where culture steps in. Adults across the globe eat very different food, and children, by and large, eat what they are given. The result is that we get used, as we grow up, to eating food that other people may think bizarre, even disgusting. Would you like to eat insect larvae, whale meat or sea cucumbers? Plenty of people do. Indeed, these foods are considered a delicacy in many countries.

There aren't just national differences in diet. In the UK, for example, there are regional differences – not just between England, Wales, Scotland and Northern Ireland but within each of these four areas too. Then there are differences related to social class (think about it!), age and gender. Marketing people know this only too well and carefully target food advertisements at the right 'segment' of the population.

And yet, one of the interesting things about what we eat nowadays is the result of globalisation. Most of us eat a greater variety of foods today than our grandparents did. You probably eat Indian, Chinese and Italian foods, to mention just three nationalities. All in all we are remarkably adaptable as far as our diet goes.

Carbohydrates

The term carbohydrate was first used in the nineteenth century and means 'hydrated carbon'. If you look at each carbon in a carbohydrate molecule (see Figure 1.29), you should be able to work out why, bearing in mind that hydration means adding water.

Most people are familiar with sugar and starch being classified as carbohydrates, but the term covers a large group of compounds with the general formula $C_x(H_2O)_n$.

Sugars are either **monosaccharides**, single sugar units, or **disaccharides**, in which two single sugar units have combined in a condensation reaction. See Figures 1.28 and 1.29. Long straight or branched chains of sugar units form **polysaccharides**.

Activity

Complete the interactive tutorial in **Activity 1.14** to help you understand carbohydrate structure. **A1.14S**

Monosaccharides can be joined by condensation reactions to form **disaccharides** and **polysaccharides** containing three or more sugar units.

Figure 1.28 A simplified diagram to show how simple sugar units (monomers) can be joined to form more complex carbohydrates (polymers).

Key biological principle: Large biological molecules are often built from simple subunits

All organisms rely on the same basic building blocks as a result of our shared evolutionary origins. Hydrogen, carbon, oxygen and nitrogen account for more than 99% of the atoms found in living organisms. Relatively simple molecules join together in different ways to produce many of the large important biological molecules.

Polymers, such as polysaccharides (Figure 1.28), proteins and nucleic acids, are made by linking identical or similar subunits, called monomers, to form straight or branched chains. Lipids are another group of biological molecules

also constructed by joining smaller molecules together, though they are not polymers since they are not chains of monomers. Large biological molecules have structures that are well suited to their functions.

In Topic 1 we are looking at the structure and function of some carbohydrates and lipids, returning in later topics to see how these molecules have many other roles. In Topic 2 the structure and function of nucleic acids and proteins will be examined in detail.

Monosaccharides

Monosaccharides are single sugar units with the general formula $(CH_2O)_n$, where n is the number of carbon atoms in the molecule. Monosaccharides have between three and seven carbon atoms, but the most common number is six. For example, the monosaccharides **glucose**, **galactose** and **fructose** all contain six carbon atoms and are known as **hexose** sugars (Figure 1.29).

Q1.20 What are the ratios of carbon, hydrogen and oxygen in monosaccharides?

A hexose sugar molecule has a ring structure formed by five carbons and an oxygen atom; the sixth carbon projects above or below the ring. The carbon atoms in the molecule are numbered, starting with 1 on the extreme right of the molecule. The side branches project above or below the ring, and their position determines the type of sugar molecule and its properties.

glucose

Glucose is important as the main sugar used by all cells in respiration. Starch and glycogen are polymers made up of glucose subunits joined together. When starch or glycogen is digested, glucose is produced. This can be absorbed and transported in the bloodstream to cells.

this can also be shown as:

galactose

Galactose occurs in our diet mainly as part of the disaccharide sugar lactose, which is found in milk. Notice that the –OH groups on carbon 1 and carbon 4 lie on the opposite side of the ring compared with their position in glucose.

fructose

Fructose is a sugar which occurs naturally in fruit, honey and some vegetables. Its sweetness attracts animals to eat the fruits and so help with seed dispersal.

Figure 1.29 Glucose, galactose and fructose are examples of monosaccharides. They are all hexose sugars.

Monosaccharides provide a rapid source of energy. They are readily absorbed and require little or, in the case of glucose, no change before being used in cellular respiration. Glucose and fructose are found naturally in fruit, vegetables and honey; they are both used extensively in cakes, biscuits and other prepared foods.

Q1.21 At a glance galactose and glucose look similar. Compare their molecular structures shown in Figure 1.29 and describe how they differ.

Disaccharides

Two single sugar units can join together and form a disaccharide (double sugar) in a **condensation** reaction. A condensation reaction is so called because a water molecule is released as the two sugar molecules combine in the reaction. Condensation reactions are common in the formation of complex molecules. Figure 1.30 shows the formation of the disaccharide maltose by a condensation reaction between two glucose molecules. The bond that forms between the two glucoses is known as a glycosidic bond or link.

The bond in maltose is known as a 1,4 glycosidic bond because it forms between carbon 1 on one molecule and carbon 4 on the other.

Figure 1.30 Two glucose molecules may join in a condensation reaction to form the disaccharide maltose. A water molecule is released during the reaction.

Common disaccharides found in food are **sucrose**, **maltose** and **lactose**. Their structures are shown in Figure 1.31.

Sucrose
Sucrose, formed from glucose and fructose, is the usual form in which sugar is transported around the plant.

Maltose
Maltose, formed from two glucose molecules, is the dsaccharide produced when amylase breaks down starch. It is found in germinating seeds such as barley as they break down their starch stores to use for food.

Lactose
Galactose and glucose make up lactose, the sugar found in milk.

Figure 1.31 Disaccharides formed by joining two monosaccharide units.

Q1.22 Name the monosaccharides that make up the following disaccharides and identify the glycosidic bond in each molecule shown in Figure 1.31.

a sucrose

b maltose

c lactose

The white or brown crystalline sugar we use in cooking, and also in golden syrup or molasses, is sucrose, extracted from sugar cane or sugar beet.

The glycosidic link between two sugar units in a disaccharide can be split by **hydrolysis**. This is the reverse of condensation: water is added to the bond and the molecule splits into two (Figure 1.32). Hydrolysis of carbohydrates takes place when carbohydrates are digested in the gut, and when carbohydrate stores in a cell are broken down to release sugars.

Q1.12 Using the molecule in Figure 1.31, sketch what the monomers that make up lactose would look like after hydrolysis.

If monosaccharides are eaten they are rapidly absorbed into the blood causing a sharp rise in blood sugar. Polysaccharides and disaccharides have to be digested into monosaccharides before being absorbed. This takes some time and monosaccharides are released slowly so eating complex carbohydrates does not cause swings in blood sugar levels as does eating monosaccharides.

Lactose is the sugar present in milk. Many adults are intolerant of lactose and drinking milk will produce unpleasant digestive problems for these people. One solution is to hydrolyse the lactose in milk, which converts the *disaccharide* lactose into the *monosaccharides* glucose and galactose.

Industrially this is carried out using the enzyme lactase. Lactase can be immobilised in a gel, and milk is poured in a continuous stream through a column containing beads of the immobilised enzyme (Figure 1.33). Asian and Afro-Caribbean people have a particularly high rate of lactose intolerance, so the resulting lactose-free milk is particularly suitable for food-aid programs. Untreated milk would cause further problems for people already suffering from malnutrition and dehydration.

Figure 1.32 The glycosidic bond between the two glucose molecules in maltose can be split by hydrolysis. In this reaction water is added.

Figure 1.33 Whey waste from cheese-making contains lactose. Hydrolysis of the waste produces syrup which is used in the food industry.

Activity

In **Activity 1.15** you immobilise lactase and use it to hydrolyse lactose. **A1.15**

Polysaccharides

Polysaccharides are polymers made up from simple sugar monomers joined by glycosidic links into long chains, as shown in Figure 1.34.

Figure 1.34 Glycosidic links join the glucose molecules that make up this polysaccharide.

There are three main types of polysaccharide found in food: **starch** and **cellulose** in plants, and **glycogen** in animals. Although all three are polymers of glucose molecules, they are sparingly soluble (they do not dissolve easily) and do not taste sweet.

Starch and glycogen act as energy storage molecules within cells. These polysaccharides are suitable for storage because they are compact molecules with low solubility in water. This means that they do not affect the concentration of water in the cytoplasm and so do not affect movement of water into or out of the cell by osmosis. See Topic 2 for details of osmosis.

Starch, the storage carbohydrate found in plants, is made up of a mixture of two molecules, **amylose** and **amylopectin**.

- Amylose is composed of a straight chain of between 200 and 5000 glucose molecules with 1,4 glycosidic links between adjacent glucose molecules. The position of the bonds causes the chain to coil into a spiral shape.

- Amylopectin is also a polymer of glucose but it has side branches. A 1,6 glycosidic link holds each side branch onto the main chain.

Figure 1.35 attempts to show these complex 3D structures. Starch grains in most plant species are composed of about 70–80% amylopectin and 20–30% amylose. The compact

amylose   amylopectin

Figure 1.35 The two forms of starch – amylose and the branched chain amylopectin. The chains of glucose molecules coil to form a spiral. This is held in place by hydrogen bonds that form between the hydroxyl (OH) groups which project into the centre of the spiral.

spiral structure of starch and its insoluble nature make it an excellent storage molecule. It does not diffuse across cell membranes and has very little osmotic effect within the cell.

Starch is a major source of energy in our diet, and is common in many foods (Figure 1.36). It occurs naturally in fruit, vegetables and cereals, often in large amounts. The sticky gel formed when starch is mixed with water makes it a good thickening agent and it is also added to many food products as a replacement for fat.

Figure 1.36 Foods high in starch.

Bacteria, fungi and animals store **glycogen** instead of starch. Glycogen is another polymer composed of glucose molecules. Its numerous side branches (Figure 1.37) mean that it can be rapidly hydrolysed, giving easy access to stored energy. In humans glycogen is stored in the liver and muscles.

glycogen

Figure 1.37 Glycogen, the storage carbohydrate found in animal cells, has a branched structure similar to amylopectin.

Cellulose in the diet is known as **dietary fibre**, and it is also referred to as a non-starch polysaccharide. Up to 10 000 glucose molecules are joined to form a straight chain with no branches (the glucose molecules have a slightly different structure to those found in starch). The structure of cellulose is considered in Topic 4; it is not required at this stage.

Indigestible in the human gut, cellulose has an important function in the movement of material through the digestive tract. Dietary fibre is thought to be important in the prevention of 'Western diseases' such as coronary heart disease, diabetes and bowel cancer.

Lipids

Lipids enhance the flavour and palatability of food, making it feel smoother and creamier (Figure 1.38). They supply over twice the energy of carbohydrates, 37 kJ of energy per gram of food, and you will see how in Topic 7. This can be an advantage if large amounts of energy need to be consumed in a small mass of food. It also means a large amount of energy can be stored in a small mass, for example in seeds.

Lipids are organic molecules found in every type of cell; they are insoluble in water but soluble in organic solvents such as ethanol. The most common

Figure 1.38 Which is more popular – with or without fat?

lipids that we eat are **triglycerides**, used as energy stores in plants and animals. Triglycerides are made up of three fatty acids and one glycerol molecule linked by condensation reactions (Figure 1.39). The bond that forms between each fatty acid and the glycerol is known as an ester bond. Three ester bonds are formed in a triglyceride.

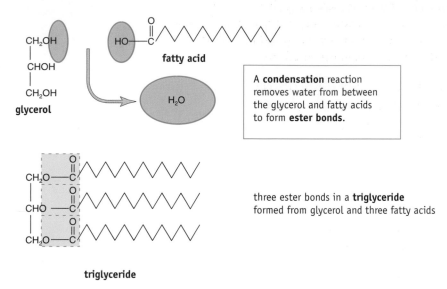

A **condensation** reaction removes water from between the glycerol and fatty acids to form **ester bonds**.

three ester bonds in a **triglyceride** formed from glycerol and three fatty acids

Figure 1.39 The formation of a triglyceride, a common type of lipid.

Saturated fats

If the fatty acid chains in a lipid contain the maximum number of hydrogen atoms, they are said to be **saturated**. In a saturated fatty acid the hydrocarbon chain is long and straight (Figure 1.40).

There are no carbon to carbon double bonds in the saturated fatty acid chain and no more hydrogens can be added to it. Animal fats from meat and dairy products are major sources of saturated fats.

Straight, saturated hydrocarbon chains can pack together closely. The strong intermolecular bonds between triglycerides made up of saturated fatty acids result in fats that are solid at room temperature.

Figure 1.40 A A saturated hydrocarbon chain. **B** Palmitic acid, a saturated fatty acid with a straight hydrocarbon chain containing 16 carbons in total. It can be drawn in several ways.

Unsaturated fats

Monounsaturated fats have one double bond between two of the carbon atoms in each fatty acid chain (Figure 1.41). **Polyunsaturated** fats have a larger number of double bonds. A double bond causes a kink in the hydrocarbon chain; these kinks prevent the unsaturated hydrocarbon chains packing closely together. The weaker intermolecular bonds between unsaturated triglycerides result in oils that are liquid at room temperature. Olive oil is particularly high in monounsaturated fats. Most other vegetable oils, nuts and fish are good sources of polyunsaturated fats.

Figure 1.41 A An unsaturated hydrocarbon chain. **B** Oleic acid has a double bond which causes a kink in the hydrocarbon chain. It can be drawn in several ways.

Other types of lipid

Cholesterol (Figure 1.42) is a short lipid molecule. It is a vital component of cell membranes with roles in their organisation and functioning. The steroid sex hormones (such as progesterone and testosterone) and some growth hormones are produced from cholesterol. Bile salts, involved in lipid digestion and assimilation, are formed from cholesterol. For all these reasons, cholesterol is essential for good health. Cholesterol is made in the liver from saturated fats and also obtained in our diet, found associated with saturated fats in foods such as eggs, meat and dairy products. However, there are concerns that too high a blood cholesterol level can be bad for us.

Figure 1.42 Alternative ways of showing the structure of cholesterol.

Phospholipids are similar to triglycerides but one of the fatty acids is replaced by a negatively charged phosphate group. Phospholipids are important components of cell membranes, which you will study in Topic 2.

As well as supplying energy in the diet, fats also provide a source of **essential fatty acids**, that is, fatty acids that the body needs but cannot synthesise. In addition, fat-soluble vitamins (A, D, E and K) can only enter our diet dissolved in fats. Fats must therefore be present in a balanced diet to avoid deficiency symptoms. For example, a deficiency of linoleic acid (an essential fatty acid) can result in scaly skin, hair loss and slow wound healing.

The energy balance

Look on food labels and you often see recommended daily amounts for nutrients, along with daily energy requirements for men and women. But how much energy is right for each of us, and what happens if we don't get it right?

Getting it right

The UK Department of Health publishes dietary guidelines for most nutrients. They used to give recommended daily amounts but in 1991 these were largely replaced with dietary reference values (DRVs). DRVs are estimates of the requirements and are not recommendations or goals for individuals. These include:

- an estimated average requirement (EAR)
- a lower reference nutrient intake (LRNI)
- a higher reference nutrient intake (HRNI).

These effectively provide a range of values within which a healthy balanced diet should fall (Figure 1.43). Upper and lower limits have not been set for carbohydrates and fats. Instead, estimated average requirements for energy are suggested plus the average percentage that should come from the different energy components of a diet. Tables 1.4 and 1.5 give the recommendations.

Q1.24 Look at Table 1.4. Compare the energy requirements of males and females, commenting on changes with age.

Activity

Activity 1.16 Complete this ICT-based tutorial to understand lipid structure. **A1.16S**

Figure 1.43 For a group of people who receive adequate nutrients, the range of intakes will vary around the EAR.

Estimated average requirements (EARs) for energy kJ/day (kcalories/day)				
Age/years	**Males**		**Females**	
1–3	5150	1230	4860	1165
4–6	7160	1715	6460	1545
7–10	8240	1970	7280	1740
11–14	9270	2220	7720	1845
15–18	11510	2755	8830	2110
19–50	10600	2550	8100	1940
51–59	10600	2550	8000	1900
60–64	9930	2380	7990	1900
65–74	9710	2330	7960	1900
74+	8770	2100	7610	1810

Table 1.4 Estimated average requirements for energy in kJ per day.

Weblink

The British Nutrition Foundation website provides detailed information on nutrient requirements and recommendations.

% of daily total food energy intake excluding alcohol			
Fat		Carbohydrates	
Saturated	Unsaturated	Starch	Sugars
11	24	37	13

Table 1.5 Dietary guidelines for percentage of daily energy that should come from carbohydrate and fat.

Activity

Activity 1.17 uses dietary analysis software to help you to work out your own energy budget and determine whether you are getting the right amount of energy from the best sources. **A1.17S**

Getting it wrong

We have to be aware that we need both carbohydrates and fats in our diet for good health, but that there are consequences if we get it wrong by consuming too much energy or if the percentage supplied by the various components differs greatly from the guidelines.

You need a constant supply of energy to maintain your essential body processes, such as the pumping of the heart, breathing and maintaining a constant body temperature. These processes go on all the time, even when you are completely 'at rest'. The energy needed for these essential processes is called the **basal metabolic rate** (BMR) and varies between individuals. BMR is higher in:

- males
- heavier people
- younger people
- more active people.

Q1.25 The estimated average requirements for energy shown in Table 1.4 were calculated by multiplying basal metabolic rate by a physical activity level factor of 1.4, which reflects current levels of physical activity. This gives estimates suitable for people who do little physical activity at work or in leisure time. How would this calculation be altered to give estimates for more active people?

The 'average' person may require between 8000 and 10 000 kJ a day but an athlete may require double this quantity. Some cyclists consume four times this amount.

If you eat fewer kilojoules per day than you use, you have a negative energy balance and energy stored in the body will be used to meet the demand. A regular shortfall in energy intake will result in weight loss. If you routinely eat more energy than you use, you have a positive energy balance. The additional energy will be stored and you will put on weight (Figure 1.44).

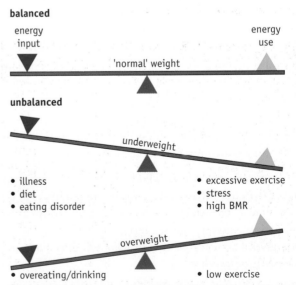

Figure 1.44 The balance between energy input and energy output determines whether the body maintains, gains or loses weight.

In the UK, it is estimated that over 60% of men (16 years and over) and over 50% of women (16 years and over) are either overweight or obese. Approximately 20% of the adult population are obese, and the rate of **obesity** in men has tripled since the mid 1980s. The 2003 Health Survey of England reported that 27.7% of 2–10-year-olds were overweight, including 13.7% who were obese. The increasing prevalence of obesity among both children and adults has been called an epidemic in Western Europe and the USA.

Defining 'overweight' and 'obese'

Body mass index (BMI) is a conventionally used method of classifying body weight relative to a person's height. To calculate BMI, body mass (in kg) is divided by height (in metres) squared, i.e.

$$BMI = \frac{body\ mass/kg}{height^2/m^2}$$

For example, the BMI of a person with a body mass of 65 kg and height of 1.72 m is 22.0. This figure can then be used to identify the category of body weight to which that person belongs, as shown in Table 1.6. BMI does not have an exact correlation with fat levels in the body. BMI may not be accurate for athletes, people over 60, or those with long-term health conditions.

BMI	Classification of body weight
<20	Underweight
20–24.9	Normal
25–29.9	Overweight
30–40	Obese
>40	Severely obese

Table 1.6 The use of BMI to classify body weight.

Q1.26   Calculate the body mass index of a person with a body mass of 85 kg and height of 1.68 m. How would you describe the body weight of this person?

There is evidence that **waist-to-hip ratio** is a better measure of obesity than BMI and shows a highly significant association with risk of heart attack. A large scale case-control study looked at 27 098 people in 52 countries. 12 461 of these people had already had a heart attack and they were matched with control individuals of the same sex and similar age who had no history of cardiovascular disease. The participants were asked about their health, economic status, lifestyle and family history of coronary heart disease. Their weight, height, waist and hip circumference were measured to allow BMI and waist-to-hip ratios to be calculated.

BMIs in men and women who had previously had a heart attack were only slightly higher than the BMIs of the control group. However, their waist-to-hip ratios were much higher than those of the control group. There is a continuous positive correlation between waist-to-hip ratio and heart attack (Figure 1.45). Waist-to-hip ratio gives a better indication of who is at risk of a heart attack, even in people with low BMIs of less than 20.

Waist-to-hip ratio is calculated by dividing waist circumference by hip circumference. The waist is measured unclothed at the narrowest point between the rib margin and

Figure 1.45 The relationship between waist-to-hip ratio and heart attack.

the top of the hip bone. The hip circumference is measured in light clothing at the widest point around the buttocks. A non-stretchable tape measure is used attached to a spring scale with a tension of 750 g. Ideally men should not have a waist-to-hip ratio over 0.90 and women's should not be greater than 0.85.

Q1.27 A 45-year-old man has a waist measurement of 91 cm and a hip measurement of 115 cm. Calculate his waist-to-hip ratio and comment on his risk of heart disease.

A poor diet, particularly one high in fat, and a sedentary lifestyle are the major contributing factors to the development of obesity. There is evidence in the UK that fat consumption has actually declined since 1990, but greater inactivity means that obesity and associated conditions are on the increase.

A high-fat diet will not necessarily result in weight gain if combined with a high level of physical activity. This was clearly illustrated by Ranulph Fiennes and Mike Stroud during their 1993 expedition to cross the Antarctic on foot – they found 23 000 kJ a day inadequate to meet their energy needs (Figure 1.46).

Consequences of obesity

Obesity increases your risk of coronary heart disease and stroke, even without other risk factors being present. The more excess fat you carry, especially around your middle, the greater the risk to your heart. Obesity can also greatly increase your risk of type II diabetes. Type II diabetes is also referred to as non-insulin-dependent diabetes or late-onset diabetes. It, in turn, increases your risk of coronary heart disease and stroke.

Figure 1.46 Over 23 000 kJ a day and Ranulph Fiennes still lost weight! Even with a padded jacket he looks thin.

Did you know?

What is type II diabetes?

There are over 2 million people with diabetes in the UK and probably at least another 750 000 people with diabetes who don't know they have the condition. Type I occurs when the body is unable to make insulin.

In type II diabetes the body either does not produce sufficient insulin or the body fails to respond to the insulin that is produced. You probably know that **insulin** is the hormone that helps regulate **blood glucose levels**. After a meal, the level of blood

CONTINUED ▶

glucose rises; in response to this change the **pancreas** produces insulin and secretes it into the bloodstream. The insulin causes cells to absorb glucose, thus returning the level in the blood to normal. Continually high blood glucose levels due to frequent consumption of sugar-rich foods can reduce the sensitivity of cells to insulin, and type II diabetes results. It may take years to develop and may not even be diagnosed. It is thought that up to a million people in the UK could be unaware that they have type II diabetes. Type II is the more common of the two main types and accounts for between 85 and 95% of all people with diabetes.

Obesity can also raise your blood pressure and elevate your blood lipid levels, two classic risk factors for cardiovascular disease. Studies have shown a positive correlation between the percentage of saturated fat in the diet and high blood pressure. They have also shown a positive correlation between the percentage of saturated fat in the diet and increased incidence of cardiovascular disease. Much media attention, particularly in advertising, is focused on saturated fats and cholesterol.

Why is cholesterol such a problem?

There is a considerable amount of evidence to show that the higher your blood cholesterol level, the greater your risk of coronary heart disease (Figure 1.47).

Q1.28 Using the data in Figure 1.47, comment on the relationship between serum cholesterol levels and the risk of death from coronary heart disease.

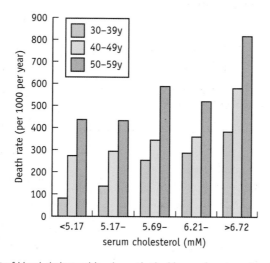

Figure 1.47 The effect of blood cholesterol levels on the incidence of coronary heart disease, data from the Framingham Study.

It is estimated that in the UK 45% of deaths from coronary heart disease in men and 47% of deaths from coronary heart disease in women are due to a raised blood cholesterol level (greater than 5.2 mmol/l). It is thought that 10% of deaths from coronary heart disease in the UK could be avoided if everyone had a blood cholesterol level of less than 6.5 mmol/l.

However, as you may realise, it's not quite as simple as that! Like all lipids, cholesterol is not soluble in water. In order to be transported in the bloodstream, insoluble cholesterol is combined with proteins to form soluble **lipoproteins**.

There are two major transport lipoproteins:

- **low-density lipoproteins (LDLs)** – the main cholesterol carrier in the blood. The triglycerides from saturated fats in our diet combine with cholesterol and protein to form LDLs. These circulate in the bloodstream and bind to receptor sites on cell membranes before being taken up by the cells. Excess LDLs in the diet overload these membrane receptors, resulting in high blood cholesterol levels. Saturated fats may also reduce the activity of LDL receptors so the LDLs are not removed from the blood, thus further increasing the blood cholesterol levels. This cholesterol may be deposited in the artery walls forming atheromas.

- **high-density lipoproteins (HDLs)** – HDLs have a higher percentage of protein compared with LDLs, hence their higher density. High-density lipoproteins are made when triglycerides from unsaturated fats combine with cholesterol and protein. HDLs transport cholesterol from the body tissues to the liver where it is broken down. This lowers blood cholesterol levels and helps remove the fatty plaques of atherosclerosis.

Monounsaturated fats are thought to help in the removal of LDLs from the blood. Polyunsaturated fats are thought to increase the activity of the LDL receptor sites so the LDLs are actively removed from the blood.

LDLs are associated with the formation of atherosclerotic plaques whereas HDLs reduce blood cholesterol deposition. Therefore, it is desirable to maintain a higher level of HDL (the so-called 'good cholesterol' or 'protective cholesterol') and a lower level of LDL (the so-called 'bad cholesterol'). Eating a low-fat diet which particularly avoids saturated fats will help reduce total blood cholesterol, and especially LDL cholesterol, which constitutes the major component of the cholesterol risk for CVD.

Activity

In **Activity 1.18** you look at some evidence for a causal relationship between blood cholesterol levels and cardiovascular disease and some conflicting evidence. **A1.18S**

Q1.29 Up until the menopause, women generally have higher HDL:LDL ratios than men. What consequences would you expect this to have for the incidence of coronary heart disease in women compared to men?

Q1.30 A person stops eating butter on their toast and starts using polyunsaturated spread instead. What effect will this have on their blood LDL levels? Explain your answer.

Measuring blood LDL level is difficult and expensive, but it is relatively easy to measure the total cholesterol level. You can even test your own cholesterol at home now, using a home test kit (Figure 1.48). Testing for total cholesterol may be part of a cardiovascular assessment for someone suspected of being at risk of CVD.

Conflicting evidence

Many studies provide evidence for a positive correlation between fat consumption and CHD mortality rates (Figure 1.47). However,

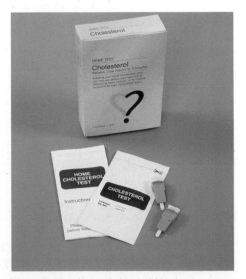

Figure 1.48 A home cholesterol kit.

conflicting evidence has also been produced. For example, in France CHD is low despite high intake of cholesterol and saturated fat. The mean total blood cholesterol levels in France and the UK are similar but the mortality rate due to CHD is much lower in France. Several hypotheses have been put forward to explain this. It has been proposed that the regular moderate consumption of alcohol in France may account for the difference in CHD mortality. Studies suggest that moderate wine drinking has a protective effect against CHD with the antioxidants in wine preventing the oxidation of low density lipoproteins which is involved in plaque formation. It may also be that the French diet contains a higher percentage of HDLs.

Smoking

Smoking cigarettes is one of the major risk factors for the development of cardiovascular disease. The constituents in smoke affect the circulatory system in the following ways:

- The haemoglobin in red blood cells carries carbon monoxide from the smoke instead of oxygen. This reduces the supply of oxygen to cells. Any narrowing of the arteries due to atherosclerosis will reduce blood flow through the arteries in the heart and brain, further reducing the oxygen supply to the cells of the heart and brain. This will result in an increased heart rate as the body reacts to provide enough oxygen for the cells.

- Nicotine in smoke stimulates the production of the hormone adrenaline. This hormone causes an increase in heart rate and also causes arteries and arterioles to constrict, both of which raise blood pressure.

- The numerous chemicals that are found in smoke can cause damage to the lining of the arteries, triggering atherosclerosis.

- Smoking has also been linked with a reduction in HDL cholesterol level.

Inactivity

The British Heart Foundation considers physical inactivity to be one of the most common risk factors for heart disease. They estimate that only three or four in every 10 men and two or three in every 10 women do sufficient exercise to give some protection against heart disease. It has been shown that being active can halve the risk of developing coronary heart disease.

Moderate exercise, such as walking, cycling or swimming, helps prevent high blood pressure and can help to lower it. Exercise not only helps maintain a healthy weight, it also seems to raise HDL cholesterol without affecting LDL cholesterol levels. It also reduces the chance of developing type II diabetes and helps in controlling the condition.

A person who is physically active is much more likely to survive a heart attack or stroke compared with someone who has been inactive. The Department of Health recommends at least 30 minutes of at least moderate exercise five times a week, and this could be built up in 10 minute periods.

My dad had a heart attack – will I?

If one or other of your parents suffer or suffered from cardiovascular disease, you are more likely to develop it yourself. There may be inherited predisposition for the disease.

Heredity and risk

Some diseases, such as some simple genetic disorders, have a single risk factor. These diseases are 'determined' by the inheritance of a defective allele, and the risk of suffering from the conditions follows the Mendelian rules of inheritance that you have already met at GCSE.

For example, consider two people who are carriers of sickle cell anaemia. This is a recessive genetic condition in which an abnormal form of haemoglobin, the oxygen-carrying protein found in red blood cells, is produced. This abnormal haemoglobin is less soluble and at low oxygen concentrations it crystallises, causing the red blood cells to become distorted. The sickle- or crescent-shaped cells are less efficient at carrying oxygen and can block small blood vessels.

These two people decide to try to have a child. The chances of this child, or any others they have, inheriting the defective form of the gene from both parents is 1 in 4, as shown in Figure 1.49.

Parent genotype	Aa		Aa	
Parent phenotype	normal red blood cells		normal red blood cells	
Gamete	A	a	A	a

	A	a
A	AA	Aa
a	Aa	aa

Any child with this genotype will have sickle cell anaemia.

Figure 1.49 There is a 1 in 4 chance of any child inheriting two copies of the defective allele and suffering from sickle cell anaemia.

With some genetically inherited conditions the risks are very clear cut. Often though, inheritance is much more complex. Even in some conditions that are controlled by a single gene, it is now known that different mutations of that gene determine how severe the sufferer's condition is. This is true for cystic fibrosis, which you will study in detail in Topic 2.

Some diseases result from several genes interacting. In other diseases, genes have been identified that do not *cause* the condition directly but that still increase the individual's chance of developing it. There is no clear-cut relationship between having these genes and having the condition; rather the genes *increase* the individual's susceptibility to the disease.

Genes and CHD

There are some single gene disorders that increase the likelihood of early development of coronary heart disease. For example, in familial hypercholesterolaemia (FH), mutations in the LDLR gene cause the LDL receptors, involved in removal of LDL from the blood, not to form or to have a shape that makes them less efficient. This results in high blood LDL levels with early onset of CHD. It is thought that about one in 500 people carry a mutation in this gene and it may account for 5–10% of the coronary artery disease in people below 55 years of age. However, even with this gene mutation the development of CHD will still be dependent on other environmental factors, such as lifestyle and diet.

Activity

You can read about the role of genes in the sudden death of athletes in **Activity 1.19. A1.19S**

The inheritance of cardiovascular disease is not a simple case of a single faulty gene for the condition being passed from one generation to the next. There are several genes that can affect your likelihood of developing cardiovascular disease. The apolipoprotein gene cluster has been identified as associated with coronary heart disease and other conditions such as Alzheimer's disease. This group of genes has been extensively researched, and it appears that some are linked to higher risk whereas others may reduce the risk.

Apolipoproteins are the protein component of lipoproteins. They are mostly formed in the liver and intestines and have important roles in stabilising the structure of the lipoproteins and recognising receptors involved in lipoprotein uptake on the plasma membrane of most cells in the body. There are several types of apolipoproteins, including:

- Apolipoprotein A (APOA) – the major protein in HDL, which helps in the removal of cholesterol to the liver for excretion. Mutations in the apoA gene are associated with low HDL levels and reduced removal of cholesterol from the blood, leading to increased risk of coronary heart disease.

- Apolipoprotein B (APOB) – the main protein in LDL, the molecule which transfers cholesterol from the blood to cells. Mutations of the apoB gene result in higher levels of LDL in the blood and a higher susceptibility to CVD.

- Apolipoprotein E (APOE) – a major component of HDLs and very low-density lipoproteins (VLDLs), which are also involved in removal of excess cholesterol from the blood to the liver. The apoE gene has three common alleles, producing three forms of the protein, E2, E3 and E4. APOE4 slows removal of cholesterol from the blood and therefore having the E4 allele may increase the risk of coronary heart disease.

Q1.31 How many phenotypes of APOE are possible?

Q1.32 One mutation of APOA, known as milano (because it is common in people from Milan in Italy), results in very low levels of HDL, but each HDL is very efficient at removing cholesterol from the blood. How will this affect the risk of developing CVD for those who are carriers of the mutation?

There are numerous mutations in the APO genes and the effects of these mutations are modified in different environments and when in different combinations, making it impossible to estimate the effect of a single gene or single environmental risk factor. In the future, testing for levels of apolipoproteins in a person's blood may be a better indicator of a person's risk of CHD than cholesterol levels themselves.

However, the chance that a person will suffer from cancer or cardiovascular disease, for example, is rarely the consequence of genetic inheritance alone. Many such diseases are **multifactorial**, with heredity, the physical environment, the social environment and lifestyle behaviour choices *all* contributing to the risk. The combination of risk factors experienced by the individual determines their risk of developing the disease.

Families do not pass on just their genes. You may also acquire your parents' lifestyle and its associated risk factors, such as smoking, lack of exercise and poor diet.

Figure 1.50 Genetic testing is becoming more common. Some companies now provide such testing so that individuals can see if they are at high or low risk of developing certain diseases.

Other risk factors

The most important factors that increase the risk of cardiovascular disease are smoking, having high blood pressure, having too high a level of blood cholesterol and lack of physical activity. As we have seen, obesity is also a major risk factor. In addition, there are other things you should think about when deciding what to do if you want to lower your CVD risk.

The role of antioxidants

During reactions in the body, unstable radicals result when an atom has an unpaired electron, e.g. in the superoxide radical, $O_2^{\bullet}$. Radicals (sometimes known as free radicals) are highly reactive and can damage many cell components including enzymes and genetic material. This type of cellular damage has been implicated in the development of some types of cancer, heart disease and premature ageing. Some vitamins, including vitamin C, beta-carotene and vitamin E, can protect against radical damage. They provide hydrogen atoms that stabilise the radical by pairing up with its unpaired electron. The MONICA study (mentioned on page 22) found that high levels of antioxidants seemed to protect against heart disease. Partly because of their function as a good source of antioxidants, the current recommendation is to include five portions of fruit or vegetables per day in our diet.

Wine and some fruit juices contain chemicals which have antioxidant properties and also help stop platelets sticking together.

Q1.33 How might the antioxidants in wine help reduce the incidence of cardiovascular disease?

Salt

You need some salt in your diet for the healthy functioning of your body but too much can have adverse effects. The Food Standards Agency recommends a salt intake of no more than 6 g per day for an adult, but the UK average intake is double that figure. Approximately 75% of the salt eaten comes from processed food. An average-sized bowl of cereal can contain as much as 1 g of salt; a standard-sized bag of crisps about 0.5 g. Often the value for sodium rather than salt is given; the value for salt is about 2½ times greater. A high-salt diet causes the kidneys to retain water. Higher fluid levels in the blood result in elevated blood pressure with the associated cardiovascular disease risks.

Stress

How you respond to stress in your life is very important. There is evidence that coronary heart disease is sometimes linked to poor stress management. In stressful situations the release of adrenaline causes arteries and arterioles to constrict, resulting in raised blood pressure. Stress can also lead to overeating, a poor diet and higher alcohol consumption, which are all potential contributors to CVD.

> **Activity**
>
> **Activity 1.20** lets you determine if your diet contains enough antioxidant vitamins. **A1.20S**

> **Activity**
>
> **Activity 1.21** lets you investigate the quantity of the antioxidant vitamin C in fruit juice. **A1.21S**

> **Activity**
>
> **Activity 1.22** is a teacher-led demonstration that lets you take part in an investigation of some factors that affect blood pressure and heart rate. **A1.22S**

Alcohol

Heavy drinkers are at far greater risk of heart disease and a number of other diseases. Heavy drinking raises blood pressure, contributes to obesity and can cause irregular heartbeat. However, there has also been much research and debate concerning *potential* protective effects of *moderate* drinking.

If you have a glass of wine, the alcohol it contains (1 unit or 8 g) is very quickly absorbed, 20% through the wall of the stomach and the remainder through the walls of the small intestine. (One unit of alcohol is approximately the amount that an adult eliminates from the body in one hour.)

Excess alcohol consumption can result in direct tissue damage, including damage to the liver, brain and heart. Such damage contributes to an increased risk of cardiovascular disease. The liver has many functions but two of its main ones are processing carbohydrates, fats and proteins, and detoxification, including the removal and destruction of alcohol. High levels of alcohol can damage liver cells. This impairs the ability of the liver to remove glucose and lipids from the blood. In the liver alcohol is converted into ethanal, a three-carbon carbohydrate. Most of the ethanal is used in respiration but some may end up in very low-density lipoproteins (VLDLs), increasing the risk of plaque deposition.

Figure 1.51 Many larger glasses of wine available in pubs, restaurants and wine bars contain 2–3 units of alcohol, the recommended daily limit for women to avoid health problems.

Given these harmful consequences of drinking alcohol, it seems remarkable to claim that moderate drinking may actually offer some degree of protection against cardiovascular disease. However, studies have shown a small protective effect of alcohol, in particular wine, compared with abstinence. This is thought to be because moderate alcohol consumption is correlated with higher HDL cholesterol levels (the 'good' cholesterol).

If you do drink, moderation is the key! The UK recommended limits to avoid health problems are 2–3 units per day for women, and 3–4 units per day for men, with no binge drinking. There is one unit of alcohol in half a pint of average strength beer, a small glass of table wine or a measure of spirits (Figure 1.51).

Activity

In **Activity 1.23** you can find out whether caffeine increases heart rate and blood pressure. **A1.23S**

Activity

In **Activity 1.24** you can 'Test your healthy heart IQ'. **A1.24S**

Checkpoint

1.6 Produce a concept map or table which shows the risk factors for CVD and their effects.

1.4 Reducing the risks of cardiovascular disease

The risk of cardiovascular disease can be reduced in a range of different ways, including:

- stopping smoking
- maintaining resting blood pressure below 140/85 mmHg
- maintaining low blood cholesterol level
- maintaining a normal BMI/low waist-to-hip ratio
- taking more physical exercise
- moderate or no use of alcohol.

If people in the UK did not smoke, the British Heart Foundation estimates that 10 000 fewer men and women of working age would die from heart attacks each year. After stopping smoking, your risk of coronary heart disease is almost halved after only one year.

Controlling blood pressure

If a person is diagnosed with high blood pressure, changes in diet and lifestyle would be recommended. Medications are also available to reduce high blood pressure. These would normally be prescribed for people with sustained systolic pressure of ≥160 mmHg or sustained diastolic pressure of ≥100 mmHg. People with sustained blood pressure over 140/90 mmHg and who show evidence of CVD may also be treated.

Three main types of drugs are used to treat high blood pressure.

ACE inhibitors

ACE inhibitors (angiotensin converting enzyme inhibitors) are effective antihypertensive drugs which reduce the synthesis of angiotensin II. This hormone causes vasoconstriction of blood vessels to help control blood pressure. The ACE inhibitors prevent the hormone being produced from an inactive form, angiotensin I, thus reducing vasoconstriction and lowering blood pressure.

Some people experience side effects when taking ACE inhibitors. These include a dry cough, dizziness due to rapid lowering of blood pressure, abnormal heart rhythms and a reduction in the function of the kidney. However, for anyone with kidney disease the drug may reduce the risk of kidney failure. Patients intolerant to ACE inhibitors will be prescribed an alternative, which affects the angiotensin II receptors.

Calcium channel blockers

Calcium channel blockers are antihypertensive drugs that block the calcium channels in the muscle cells in the lining of arteries. For the muscle to contract, calcium must pass through these channels into the muscle cells. Failure of calcium to enter the cell prevents contraction of the muscle, the blood vessels do not constrict, and this lowers blood pressure.

There are some side effects with these drugs, such as headaches, dizziness, swollen ankles due to a build-up of fluid in the legs, abnormal heart rhythms, flushing red in the face and constipation. In people with heart failure, taking some types of calcium channel blockers can make symptoms worse or be fatal, so they may not be prescribed to people who have had a heart attack.

Diuretics

Diuretics increase the volume of urine produced by the kidneys and thus rid the body of excess fluids and salt. This leads to a decrease in blood plasma volume and cardiac output (volume of blood expelled from the heart in a minute), which lowers blood pressure.

A few people taking diuretics may have some side effects, such as dizziness, nausea or muscle cramps.

Most people take two of these drugs to control blood pressure, usually an ACE inhibitor in combination with one of the others. In 2006 the British Hypertension Society and the National Institute for Clinical Excellence published guidelines on the appropriate selection of these drugs to use, with the drugs selected according to what is effective and carries the lowest risk of side effects.

Reducing blood cholesterol levels

There is evidence from both the UK and the USA that untargeted general population cholesterol screening combined with dietary advice has little effect on lowering blood cholesterol levels. A further problem is associated with 'labelling' people. When told that they have high blood pressure, many people react by signing off sick!

Q1.34 Why do you think untargeted screening and dietary advice are not very effective at lowering blood cholesterol?

There is a need to reduce blood cholesterol in some people. One way to achieve this is through a low-fat diet. The media constantly remind us of the need to do this.

Table 1.7 shows the extent of cholesterol-lowering that is obtained from following a low-fat diet in high-risk patients, namely people who had already experienced a heart attack, compared with the general population.

Table 1.7 The effect of lipid-lowering diets in reducing blood cholesterol levels.

	Blood cholesterol reduction/	
	mmol/l	%
general population	0.22	3
high-risk patients	0.65	9

Q1.35 Can you suggest why the effect of dietary change appears to have been more successful in reducing blood cholesterol in patients who had experienced a heart attack than in the general population?

Cholesterol-lowering drugs

Individuals diagnosed with high cholesterol may also be prescribed cholesterol-lowering drugs. The main types used are statins. Statins work by inhibiting an enzyme involved in the production of LDL cholesterol. Several large studies have investigated the effects of these drugs, and researchers have collaborated to analyse the results of 14 randomised trials. The results, drawing on 90 000 participants, were published in 2005, and showed that statins quickly reduce the risk of heart disease and stroke. Studies have shown that there is a linear relationship between absolute reductions in LDL cholesterol and reductions in the incidence of major vascular events such as heart attack and stroke. In patients who have CVD, treatment with statins can reduce the risk of heart attack by up to 33%. In people who do not have CVD but have elevated cholesterol levels, statins lower total and LDL cholesterol by more than 20% and the risk of CVD by a similar amount.

Some studies have raised concerns that the use of statins may increase the risk of death by other non-vascular causes, and also increase the risk of developing particular cancers, such as gastrointestinal and respiratory cancers. To determine if this was the case, researchers in the UK and Australia analysed the results of 14 large randomised trials for statins. They found no evidence that lowering LDL cholesterol by 1 mmol/l with five years of statin treatment increased the risk of death by other non-vascular causes or of developing particular cancers. They concluded that any risks were extremely small in relation to the benefits of treatment; statin treatment can safely reduce the five-year incidence of major vascular events by 20% per mmol/l reduction in LDL cholesterol, regardless of the lipid levels at the start of treatment. They did however highlight the need for extended study of patients using statins over long periods to identify any adverse effects that might occur with long-term use.

A diet to reduce the risk of cardiovascular disease

A diet to offer protection against cardiovascular disease would include the following key features (Figure 1.52):

- Energy balanced.

- Reduced saturated fat.

- More polyunsaturated fats.

- Reduced cholesterol.

- Reduced salt.

- More non-starch polysaccharides, such as pectins and guar gum. These polysaccharides, known as soluble fibre, have been found to lower blood cholesterol. They are found in fruit, vegetables, beans, pulses and some grains (e.g. oats). They are only partially digested, forming a gel that traps the cholesterol and prevents its absorption.

- Include oily fish. Fish such as mackerel, sardines, anchovies, salmon and trout contain *omega-3 fatty acids*, a group of polyunsaturated fatty acids with their first double bond between the third and fourth carbon atoms. These fatty acids are essential for cell functioning and have been linked to a reduction in heart disease and joint inflammation. The evidence for the importance of omega-3 fatty acids is seen in the Inuit in Greenland and the inhabitants of certain Japanese islands. They regularly eat oily fish and have very low rates of coronary heart disease.

- More fruit and vegetables. Fruits and vegetables contain antioxidants and often non-starch polysaccharides.

- Include functional foods containing sterols and stanols (Figure 1.53). These are naturally produced substances in plants, similar to cholesterol. Stanols are saturated. Both sterols and stanols compete with cholesterol during its absorption in the intestine. Unfortunately, we would have to eat vast quantities of foods such as vegetable oils and grains in order to reduce our cholesterol levels through this competition. Products have therefore been developed that can incorporate sterols and stanols into everyday foodstuffs such as margarines. Controlled clinical trials of these functional foods show promising results; the amount of cholesterol absorbed from a food with stanols is about 20% compared with 50% absorption from a food without them.

Figure 1.52 A diet to reduce the risk of developing cardiovascular disease.

Review

In **Extension 1.3** you can find out more about functional food trials. **X1.03S**

Figure 1.53 Some margarines and yoghurts include sterols and stanols to help lower cholesterol.

Anticoagulant and platelet inhibitory drug treatment

If someone has had a heart attack or stroke, or is identified as being at high risk of one, in addition to the lifestyle changes and drug treatments to reduce blood pressure and blood cholesterol, they may be given drugs to prevent formation of a blood clot in an artery.

The tendency for platelet aggregation and clotting is reduced by platelet inhibitory drugs and anticoagulant drugs. Aspirin reduces the 'stickiness' of platelets and the likelihood of clot formation. But some people are allergic to aspirin, and for others it is not effective or is only partially effective. In these cases an alternative platelet inhibitory drug, clopidogrel, may be used. A combined treatment involving a daily dose of aspirin and clopidogrel can have a dramatic effect. However, it has been shown that there is a risk of bleeding in the gastrointestinal tract with aspirin, and in studies there have been high rates of serious bleeding when aspirin is used in combination with clopidogrel. This risk of bleeding may outweigh the benefits. If a person has only a low risk of a vascular event (less than 1% per year risk of a heart attack or stroke), the risk of bleeding outweighs the benefits. The risks and benefits need to be considered for each individual patient, though guidelines suggest use when the 10-year risk of a heart attack or stroke is greater than 20%.

Warfarin is an anticoagulant drug; it affects the synthesis of clotting factors. It can be taken orally for extended periods of time to prevent clotting. The benefits may be greater than with aspirin for some patients but the risk of bleeding is higher than with aspirin.

Q1.36 **a** Reeta has a high risk of heart attack and is discussing drug treatments with her doctor. What information would they have to consider before making a decision about the best drug to use?

b Regular low doses of aspirin reduce the risk of bowel cancer. Would it be sensible for everyone to take low doses of aspirin to reduce their risk of cancer? Give a reason for your answer.

Think back to Mark and Peter. What is most surprising is that Mark was only 15 when he had his stroke. Mark had no obvious risk factors that would have alerted him to the possibility of having a stroke. He reports having taken exercise, eaten a reasonably healthy diet and having not smoked. If you read Mark's full story in Activity 1.1 at the start of the topic, you will find that he had a type of stroke in which a blood vessel supplying blood to the brain bursts.

An artery can burst due to an aneurysm, where blood builds up behind a section of artery that has narrowed as a result of atherosclerosis. However, with no risk factors and in one so young, the likelihood of atheroma deposits having built up to this extent in Mark's arteries seems unlikely. Although there was no history of stroke in Mark's family, he thinks that he inherited a gene causing him to have thin artery walls more prone to bursting.

Q1.37 What additional information would you need from Mark to determine if his stroke was due to this type of inherited condition or due to atherosclerosis?

Mark remains healthy today (March 2005) and has not had a recurrence of the problem.

Peter is lucky to be alive having had a blood pressure of 240/140 mmHg, two heart attacks and heart surgery (Figure 1.55). If you look back at his story you will recall that his father died aged 53 from a heart attack. This suggests that there may have been an inherited predisposition for the condition. Thankfully, Peter's active lifestyle will have helped him survive.

Activity

In **Activity 1.25** you can discuss how people use scientific information to reduce their risk of coronary heart disease. **A1.25S**

Extension

You can read about new treatments for coronary heart disease in **Extension 1.4**. 'New treatments for cardiovascular disease' provides a fine start. It even gives you the opportunity to observe surgery! **X1.04S**

? **Did you know?**

Haemorrhagic stroke

Blood vessels on the surface of or within the brain are susceptible to bursting resulting in a stroke. A haemorrhagic stroke occurs when a blood vessel supplying blood to the brain bursts. If the burst occurs within the brain it is known as an *intracerebral haemorrhage*, whereas bursting of a vessel on the surface causes what is known as a *subarachnoid haemorrhage*. Look at Figure 1.54 and work out why the different types of stroke were given these names. There is no need for you to remember the names, but it is worth being aware of them as doctors and medical scientists use this sort of language.

Figure 1.54 Cross-section through a human brain.

Figure 1.55 A vein taken from the patient's leg is used to bypass the sections of coronary artery that are narrowed. The photograph above shows the section of vein that has been grafted onto a patient's heart.

Q1.38 Peter has two daughters and a son. What advice would you recommend he give them?

 Activity

Use **Activity 1.26** to check your notes using the topic summary provided. **A1.26S**

Topic test

Now that you have finished Topic 1, complete the end-of-topic test before starting Topic 2.

Genes and health

Why a topic called Genes and health?

It is now recognised that our genes have a major part to play in our health. Nearly 10 000 genetic disorders have been described. Some are very minor and create little or no problem; for example, colour blindness and hairy ears are both inherited conditions. Others have more serious consequences for the affected individuals. Albinism, cystic fibrosis, Huntington's disease, sickle cell disease and haemophilia are all serious conditions caused by faulty genes.

Thankfully, most serious genetic disorders are rare. But when one does occur it has a significant effect on the person's life and on the lives of their family members. Not only must affected individuals deal with the condition itself, they may also face difficult decisions about the possible inheritance of the condition by the next generation.

Cystic fibrosis dilemma

In this topic we follow Claire and Nathan, who face a daunting dilemma – should they have a child that could have cystic fibrosis (Figure 2.1)? Claire's sister had the condition, so they think it might be passed on to any child they might have. To make the decision they need more information. If the child does inherit cystic fibrosis, how will it affect the child's life and theirs? How is cystic fibrosis inherited? What are the chances of any child they have inheriting the condition? Can genetic screening help? What treatments are available now and what might be possible in the future? These are just some of the questions that need answers if Nathan and Claire are to make an informed choice.

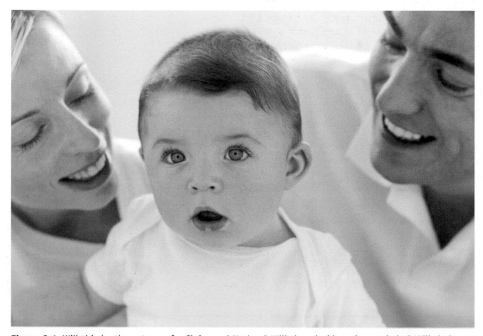

Figure 2.1 Will this be the outcome for Claire and Nathan? Will they decide to have a baby? Will their baby have CF?

Cystic fibrosis (CF) is one of the most common genetic diseases, affecting approximately 7500 people in the UK. One in 25 of the population carries the faulty CF allele. Every week on average five babies are born with CF and three young people die from it, usually as a result of lung damage. In the 1960s the average life expectancy for a child with CF was just 5 years. By the year 2000 life expectancy had risen to 31 and by 2005 to the mid 30s.

Claire is well aware of the outward symptoms of CF; she saw the problems her sister Valerie had with breathing, including a troublesome cough and repeated chest infections. Valerie also had to be very careful about her diet to ensure she overcame problems with digestion and maintained her weight. But Claire and Nathan want to know in more detail what causes these outward symptoms.

The symptoms of CF concern a sticky mucus layer that lines many of the tubes and ducts in the gas exchange, digestive and reproductive systems. The symptoms and the way CF is inherited have been known for a long time – it was first accurately described by Dorothy Andersen, a New York pathologist, in 1938. However, the genetic cause of the disease was not identified until 1989, by a group in Toronto led by Lap-Chee Tsui.

Weblink

To find out the life expectancy for someone with CF today, visit the Cystic Fibrosis Trust website.

Overview of the biological principles covered in this topic

In this topic you will study how changes in DNA can result in genetic disease using the example of cystic fibrosis. You will first look in detail at the symptoms of cystic fibrosis, extending your previous knowledge of the structure of the lungs and of gas exchange to see the importance of surface area to volume ratios in biology.

To understand the symptoms of cystic fibrosis you will study cell membrane structure and how substances move across membranes. To explain how faults arise in the cell surface membranes of a person with CF you will learn how genes code for proteins, how proteins are made and how their function is dependent on their structure. You will find out how genes play their role in inheritance.

You will discover how genetic screening and gene therapy may be used to help people with cystic fibrosis. Throughout the topic you will consider the ethical issues raised by these new technologies and you will learn how ethical arguments can be evaluated.

Activity

Activity 2.1 will give you an overview of cystic fibrosis, and let you find out more about Claire's family and the problems she and Nathan could face in the future. **A2.01S**

Read the stories in **Activity 2.2** to see how cystic fibrosis has affected some people, and how they and their families cope. **A2.02S**

Review

Are you ready to tackle Topic 2 *Genes and health?*

Complete the GCSE review and GCSE review test before you start.

2.1 The effects of CF on the lungs

The role of mucus in the lungs

The lungs allow rapid gas exchange between the atmosphere and the blood. Air is drawn into the lungs via the trachea due to low pressure in the lungs, created by the movement of the ribs and diaphragm. The trachea divides into two bronchi which carry air to and from each lung. Within each lung there is a tree-like system of tubes ending in narrow tubes, bronchioles, attached to tiny balloon-like alveoli (Figures 2.2 and 2.5). The alveoli are the sites of gas exchange.

There is nothing unusual about having a layer of **mucus** in the tubes of the gas exchange system. Everyone normally has a thin coating of mucus in these tubes, produced continuously from goblet cells in the walls of the airways. Any dust, debris or microorganisms that enter the airways become trapped in the mucus. This is continually removed by the wave-like beating of **cilia** that cover the **epithelial** cells lining the tubes of the gas exchange system. However, people with CF, like Valerie, have mucus that is drier than usual resulting in a sticky mucus layer that the cilia find difficult to move (Figure 2.3).

This sticky mucus in the lungs has two major effects on health. It increases the chances of lung infection and it makes gas exchange less efficient, particularly in the later stages of the disease.

Figure 2.3 False colour scanning electron micrograph of a clump of mucus (green/orange) surrounded by cilia (purple) in the lungs of someone with CF. Rounded goblet cells (light blue) secrete the mucus.

Figure 2.2 The fine structure of the lungs. Cilia, hair-like extensions from the surface of cells lining the trachea, bronchi and bronchioles, have an important role in keeping the lungs clean. By a repeated beating motion they move mucus and particles up and out of the lungs.

What are epithelial cells?

Epithelial cells form the outer surface of many animals including mammals. They also line the cavities and tubes within the animal, and cover the surfaces of internal organs. The cells work together as a **tissue** known as **epithelium** (plural epithelia). In Topic 3 you will study how the cells of multicellular organisms can be organised into tissues.

The epithelium consists of one or more layers of cells sitting on a **basement membrane**. This is made of protein fibres in a jelly-like matrix. There are several different types of epithelia.

The epithelium in the walls of the alveoli and capillaries is squamous (or pavement) epithelium. The very thin flattened cells fit together like crazy paving. The cells can be less than 0.2 µm thick.

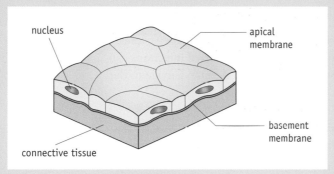

Squamous or pavement epithelium.

In the small intestine the epithelial cells extend out from the basement membrane. These column-shaped cells make up columnar epithelium. The free surface facing the intestine lumen is normally covered in microvilli, which greatly increase the surface area.

Columnar epithelium.

In the trachea, bronchi and bronchioles there are ciliated epithelial cells with cilia (hair-like structures) on the free surface. These cilia beat and move substances along the tube they line. The ciliated columnar epithelium of the gas exchange airways appears to be stratified (composed of several layers), but in fact each cell is in contact with the basement membrane. It appears to have several layers because some cells have their nucleus at the base of the cell while in others it is in the centre, giving the impression of different layers. This epithelium is therefore known as pseudostratified.

Ciliated epithelium.

CF problems

How sticky mucus increases the chances of lung infections

Microorganisms become trapped in the mucus in the lungs. Some of these can cause illness – they are **pathogens**. The mucus is normally moved by cilia into the back of the mouth cavity where it is either coughed out or swallowed, thus reducing the risk of infection. Acid in the stomach kills most microorganisms that are swallowed.

With CF the mucus layer is so sticky that cilia cannot move the mucus. Mucus production still continues, as it would in a normal lung, and the airways build up layers of thickened mucus. There are low levels of oxygen in the mucus, partly because oxygen diffuses slowly through it, and partly because the epithelial cells use up more oxygen in CF patients. Harmful bacteria can thrive in these anaerobic conditions.

Q2.1 Why might failure to move mucus create a problem?

Q2.2 Why does swallowing mucus reduce the risk of infection?

White blood cells fight the infections within the mucus but as they die they break down, releasing DNA which makes the mucus even stickier. Repeated infections can eventually weaken the body's ability to fight the pathogens, and cause damage to the structures of the gas exchange system.

How sticky mucus reduces gas exchange

Gases such as oxygen cross the walls of the alveoli into the blood system by **diffusion**. To supply enough oxygen to all the body's respiring cells, gas exchange must be rapid. The fine structure of the lungs helps to maximise this.

Key biological principle: The effect of increase in size on surface area

Living organisms have to exchange substances with their surroundings. For example, they take in oxygen and nutrients and get rid of waste materials such as carbon dioxide. In unicellular organisms the whole cell surface membrane is the exchange surface. Substances that diffuse into or out of a cell move down a **concentration gradient** (from a high to a low concentration). The gradients are maintained by the cell continuously using the substances absorbed and producing waste. For example, oxygen diffusing into a cell is used for respiration which produces carbon dioxide.

The larger an organism, the more exchange has to take place to meet the organism's needs. Larger multicellular organisms have more problems absorbing substances because of the size of the organism's surface area compared with its volume. This is known as the **surface area to volume ratio**, calculated by dividing an organism's total surface area by its volume.

Activity

Complete **Activity 2.3** to investigate the effect of surface area to volume ratio on uptake by diffusion. **A2.03S**

CONTINUED ▶

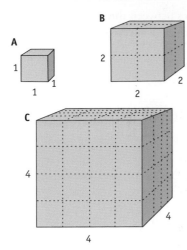

A
1
1
1

B
2
2
2

C
4
4
4

Q2.3 For each of the 'organisms' above, work out its surface area, volume and then surface area to volume ratio.

Q2.4 As the organism grows larger what happens, quantitatively, to **a** its surface area **b** its volume **c** its surface area to volume ratio?

Q2.5 Assuming that this organism relies on diffusion across its outer surface for exchange, why would it have problems if it grew any larger?

Q2.6 If you compared a tiger, a horse and a hippopotamus, which would have the smallest surface area to volume ratio?

It is clear that as organisms get larger, the surface area per unit of volume gets less. If larger organisms relied only on their general body surface for exchange of substances they could not survive because exchange would be too slow.

How can an organism increase in volume while still managing to exchange enough nutrients by diffusion?

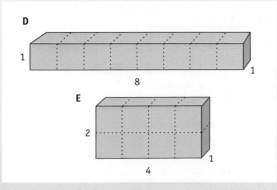

D
1
8
1

E
2
4
1

Q2.7 Work out the surface area, volume and surface area to volume ratio of organisms D and E.

Q2.8 Look at the values you have calculated for B, D and E. What do you notice?

Q2.9 If a slug, an earthworm and a tapeworm all had the same volume, which would have the largest surface area to volume ratio?

All three blocks, B, D and E, have the same volume but they have very different surface areas. The most elongated block, D, has the largest surface area to volume ratio.

Relying on the outer body surface for gas exchange is only possible in organisms with a very small volume, or in larger organisms that have a high enough surface area to volume ratio such as worms with a tubular or flattened shape.

Q2.10 Why would a land-living organism not be likely to use its entire external surface covering for gas exchange? Think of possible problems this could cause, apart from surface area to volume ratio.

Larger organisms have a variety of special organs that increase the surface area for exchange, thus increasing the surface area to volume ratio. For example, a mammal's lungs provide a large surface for gas exchange while minimising heat and water loss from the moist surface. Digestive systems provide a large surface for food absorption.

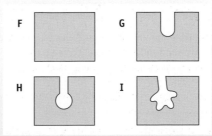

F
G
H
I

Q2.11 Which of the four organisms above, all with the same volume, has the largest surface area for exchange?

Q2.12 Name two organs that enable rapid exchange.

Q2.13 In humans, how do the substances that are absorbed get to all the distant parts of the body?

Gas exchange surfaces

Within the lungs, alveoli provide a large surface area for exchange of gases between the air and the blood. Look at Figure 2.4 and identify features of the **gas exchange surface** that you think would ensure rapid exchange between air in the alveoli and the blood.

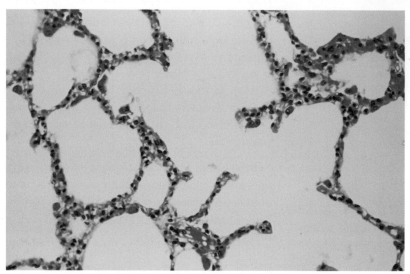

Figure 2.4 Ventilation of the lungs ensures that the air in the alveoli is frequently refreshed. This helps maintain a steep concentration gradient and maximise gas exchange across the walls of the alveoli.

You should have noticed some of these features of the gas exchange surface:

- large surface area of the alveoli
- numerous capillaries around the alveoli
- thin walls of the alveoli and capillaries meaning a short distance between the alveolar air and blood in the capillaries.

Activity

In **Activity 2.4** you will examine slides of the alveoli to observe the features that aid diffusion into and out of the bloodstream. **A2.04S**

In the web tutorial, **Activity 2.5**, you can investigate the surface area of the lungs. **A2.05S**

The body's demand for oxygen is enormous, so diffusion across the alveolar wall needs to be rapid. The rate of diffusion is dependent on three properties of gas exchange surfaces:

- **Surface area** – rate of diffusion is directly proportional to the surface area. As the surface area increases the rate of diffusion increases.
- **Concentration gradient** – rate of diffusion is directly proportional to the difference in concentration across the gas exchange surface. The greater the concentration gradient the faster the diffusion.

- **Thickness of the gas exchange surface** – rate of diffusion is inversely proportional to the thickness of the gas exchange surface. The thicker the surface the slower the diffusion.

$$\text{rate of diffusion} \propto \frac{\text{surface area} \times \text{difference in concentration}}{\text{thickness of the gas exchange surface}}$$

This is known as **Fick's law**.

The large surface area of the alveoli, the steep concentration gradient between the alveolar air and the blood (maintained by ventilation of the alveoli) and the thin walls of the alveoli and capillaries combine to ensure rapid diffusion across the gas exchange surface (Figure 2.5).

Checkpoint

2.1 Describe the properties of gas exchange surfaces.

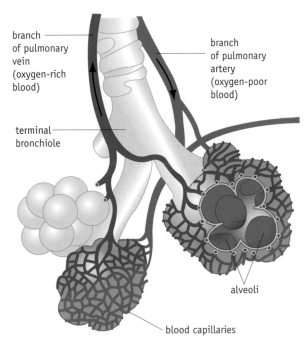

branch of pulmonary vein (oxygen-rich blood)

branch of pulmonary artery (oxygen-poor blood)

terminal bronchiole

alveoli

blood capillaries

Figure 2.5 The structure of the alveoli and the surrounding capillaries ensures that there is rapid diffusion across the gas exchange surface.

How sticky mucus might affect gas exchange

The sticky mucus layer in the bronchioles of a person with cystic fibrosis tends to block these narrow airways, preventing ventilation of the alveoli below the blockage. This reduces the number of alveoli providing surface area for gas exchange. Blockages are more likely at the narrow ends of the airways. These blockages will often allow air to pass when the person breathes in but not when they breathe out, resulting in over-inflation of the lung tissue beyond the blockage. This can damage the elasticity of the lungs.

People with CF find it difficult to take part in physical exercise because their gas exchange system cannot deliver enough oxygen to their muscle cells. The oxygen is needed for the chemical processes of aerobic respiration, which release the energy used to drive the contraction of the muscles during exercise. People with CF become short of breath when taking exercise but exercise is very beneficial to them.

2.2 Why is CF mucus so sticky?

In people with CF, the mucus layer on the surface of the epithelial cells is sticky because it contains less water than normal. The reduced water level is due to abnormal salt and water transport across the cell surface membranes caused by a faulty transport protein channel in the membrane. To understand what is going on you need to be clear about the structure and functions of proteins, the structure of cell surface membranes and how substances are transported in and out of cells. You also need to be aware of how proteins are made so you can understand what has gone wrong with the protein channel.

Key biological principle: Protein structure is the key to protein function

Proteins have a wide range of functions in living things. Antibodies, enzymes and many hormones are all protein molecules. Various proteins make up muscles, ligaments, tendons and hair. Proteins are also components of cell membranes and have important functions within the membrane. All proteins are composed of the same basic units: amino acids. There are 20 different amino acids that occur commonly in proteins. Plants can make all these amino acids whereas animals can only make some, obtaining the others through their diet. Those amino acids that animals have to obtain in their diet are known as essential amino acids.

Figure 2.6A shows the general structure of an amino acid. All amino acids contain an amine group, $-NH_2$, a carboxylic acid group, $-COOH$, and a hydrogen, $-H$, attached to a central carbon atom. Each type of amino acid has a different side chain, called a residual or R group. Glycine is the simplest amino acid with a hydrogen, $-H$, forming its R group. Alanine's R group is a $-CH_3$ group. Some amino acids have R groups with more complex carbon ring structures.

Figure 2.6 A The general structure of amino acids. **B** The structure of glycine.

Primary structure

Two amino acids join in a condensation reaction to form a dipeptide, with a peptide bond forming between the two subunits, as shown in Figure 2.7. This process can be repeated to form polypeptide chains which may contain thousands of amino acids. A protein is made up of one or more of these polypeptide chains. The sequence of amino acids in the polypeptide chains is known as the **primary structure** of a protein.

Figure 2.7 Two amino acids link in a condensation reaction to form a dipeptide. Numerous amino acids link together in this way to form a polypeptide.

Further levels of protein structure

Interactions between the amino acids in the polypeptide chain cause the chain to twist and fold into a three-dimensional shape. Lengths of the chain may first coil into α-helices or come together in β-pleated sheets (Figure 2.8). These are known as the **secondary structure**. The chain then folds into its final three-dimensional shape, the **tertiary structure** (Figure 2.9).

CONTINUED ▶

Secondary structure

The chain of amino acids may twist to form an α-helix, a shape like an extended spring (Figure 2.8A). Within the helix, hydrogen bonds form between the C=O of the carboxylic acid and the –NH of the amine group of different amino acids, stabilising the shape.

Several chains may link together, with hydrogen bonds holding the parallel chains in an arrangement known as a β-pleated sheet (Figure 2.8B). Within one protein molecule there may be sections with α-helices and other sections that contain β-pleated sheets.

A

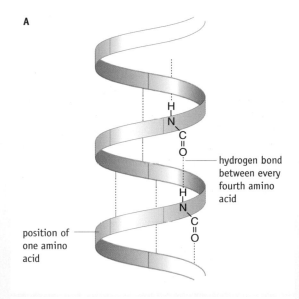

hydrogen bond between every fourth amino acid

position of one amino acid

B

hydrogen bond

Figure 2.8 The secondary structure of proteins: **A** A tightly wound α-helix. **B** A β-pleated sheet (R groups are not shown).

Support

To find out more about hydrogen bonding, have a look at the Biochemistry support on the website.

Tertiary and quaternary structure

A polypeptide chain often bends and folds to produce a precise three-dimensional shape. Chemical bonds and hydrophobic interactions between R groups maintain this final tertiary structure of the protein (Figure 2.9).

An R group is **polar** when the sharing of the electrons within it is not quite even. Polar R groups attract other polar molecules, like water, and are therefore **hydrophilic** (water-attracting). The non-polar groups are **hydrophobic** (water-repelling). Non-polar, hydrophobic R groups are arranged so they face the inside of the protein, excluding water from the centre of the molecule.

ionic bonds between ionised R groups

disulphide bond between R groups containing –SH groups

amino acids with hydrophobic side chains

amino acids with hydrophilic side chains

hydrogen bonds

Figure 2.9 The three-dimensional structure of a protein is held in place by chemical bonds between individual amino acids and by hydrophobic interactions between R groups.

A protein may be made up of several polypeptide chains held together. For example, haemoglobin, the protein found in red blood cells that carries oxygen, is made up of four polypeptide chains held tightly together in a structure known as the quaternary structure. Only proteins with several polypeptide chains have a quaternary structure; single chain proteins stop at the tertiary level.

Conjugated proteins

Some proteins are known as conjugated proteins – they have another chemical group associated with their

Activity

The interactive tutorial in **Activity 2.6** lets you review how amino acids join to form a polypeptide and then fold to achieve their three-dimensional structure. **A2.06S**

CONTINUED ▶

polypeptide chain(s). For example, in Figure 2.10 (see overleaf) the polypeptide chain that makes up myoglobin is associated with an iron-containing group.

Globular and fibrous proteins

Proteins can be divided into two distinct groups:

- globular proteins
- fibrous proteins.

In **globular proteins** the polypeptide chain is folded into a compact spherical shape. These proteins are soluble due to the hydrophilic side chains that project from the outside of the molecules and are therefore important in metabolic reactions. Enzymes are globular proteins. Their three-dimensional shape is crucial to their ability to form enzyme-substrate complexes and catalyse reactions within cells.

The three-dimensional shapes of globular proteins are critical to their roles in binding to other substances. Examples include transport proteins within membranes

and the oxygen-transport pigments haemoglobin (in red blood cells) and myoglobin (in muscle cells) (Figure 2.10). Antibodies are also globular and rely on their precise shapes to bind to the microorganisms that enter our bodies.

Fibrous proteins do not fold up into a ball shape but remain as long chains. Several polypeptide chains can be cross-linked for additional strength. These insoluble proteins are important structural molecules. Keratin in hair and skin, and collagen (Figure 2.11) in the skin, tendons, bones, cartilage and blood vessel walls, are examples of fibrous proteins.

three polypeptide chains making a strand of collagen

cross-linked strands

Figure 2.11 Collagen is a fibrous protein. Three polypeptide chains wind around each other to form a rope-like strand held together by hydrogen bonds between the chains. Each strand cross-links to other strands to produce a molecule with tremendous strength. Notice that the strands are staggered, avoiding the creation of any weak points along the length of the molecule.

Figure 2.10 The globular protein myoglobin acts as an oxygen-storage molecule in muscle cells. The oxygen (red circle) attaches to the iron within the haem group (shown green). Because the protein is associated with another chemical group it is called a conjugated protein.

Checkpoint ✔

2.2 Compare and contrast the structures and functions of globular and fibrous proteins.

Cell membrane structure

A phospholipid bilayer

CF is caused by a faulty transport protein in the surface membranes of epithelial cells. But how do we know what the structure of the cell surface membrane is like? It is so thin that under a light microscope a cross-section of the membrane looks like a single line. However, closer examination using an electron microscope reveals that it is in fact a **bilayer**, about 7 nm wide, appearing as two distinct lines (Figure 2.12). What do we mean by a bilayer? The basic structure is two layers of **phospholipids**.

Figure 2.12 The cell surface membrane appears as two distinct layers when viewed with an electron microscope. The same type of membrane surrounds many cell organelles, which you will study in Topic 3.

Look back at Figure 1.39 (page 36) and remind yourself of the structure of a lipid molecule – three fatty acids and a glycerol. Compare this with Figure 2.13 and notice the difference. In a phospholipid there are only two fatty acids; a negatively charged phosphate group replaces the third fatty acid.

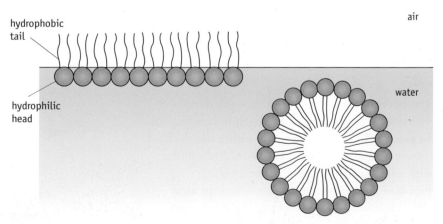

Figure 2.13 The phospholipid molecule has two distinct sections: the hydrophilic head and the hydrophobic tails. X can be a variety of chemical groups. This can be simply drawn as ○⊂.

The phosphate head of the molecule is polar; one end is slightly positive and the rest is slightly negative. This makes the phosphate head attract other polar molecules, like water, and it is therefore hydrophilic (water-attracting). The fatty acid tails are non-polar and therefore hydrophobic (water-repelling). We all know that fats and water don't mix, so when added to water phospholipids arrange themselves to avoid contact between the hydrophobic tails and the water. They may form a layer on the surface with their hydrophobic tails directed out of the water, arrange themselves into spherical clusters called micelles (Figure 2.14) or form a bilayer. The bilayer is favoured by phospholipids because the two fatty acids are too bulky to fit into the interior of a micelle. The formation of bilayers by phospholipids is of critical biological importance. A lipid bilayer will tend to close on itself so that there are no edges with exposed hydrocarbon chains, thus forming compartments. Imagine an inflated beach ball which has a double layer wall; this is like the bilayer. If you fill the inside with water, and immerse the whole thing in water, the water does not get between the two layers of the wall.

Support

To find out more about polar molecules and polar bonds, have a look at the Biochemistry support on the website.

Figure 2.14 Phospholipids in water may form a monolayer on the surface or spherical micelles.

Cells are filled with a watery or aqueous cytoplasm and are surrounded by aqueous tissue fluid. The cell surface membrane phospholipids tend to adopt their most stable arrangement themselves, which is a bilayer (Figure 2.15). This arrangement avoids the hydrophobic fatty acid tails having any contact with the water on either side of the membrane but ensures the hydrophilic phosphate heads are in contact with the water.

The fluid mosaic model

The cell surface membrane is not simply a phospholipid bilayer. It also contains proteins, cholesterol, **glycoproteins** (protein molecules with polysaccharides attached) and **glycolipids** (lipid molecules with polysaccharides attached). Some of the proteins span the membrane. Other proteins are found only within the inner layer or only within the outer layer (Figure 2.15). Membrane proteins have hydrophobic areas and these are positioned within the membrane bilayer.

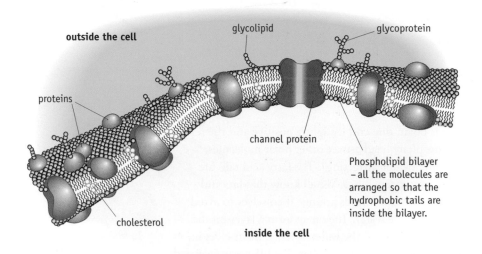

Figure 2.15 Diagram of the fluid mosaic model of the cell surface membrane.

It is thought that some of the proteins are fixed within the membrane, but others are not and can move around in the fluid phospholipid bilayer. This arrangement is known as the **fluid mosaic model** of membrane structure.

Two Californian biochemists, S. Jonathan Singer and Garth Nicholson, first proposed the fluid mosaic model in 1972. They did not carry out all the experiments that supported the model themselves, but were able to bring together evidence from a number of different research groups and incorporate it into their proposal.

Evidence for the fluid mosaic model

The most widely accepted membrane model up until the early 1970s was a three-layer protein-lipid sandwich, based on the evidence of the electron micrographs in which the dark outer layers were thought to be proteins and the lighter region within was the lipid (Figure 2.16 A). However, the protein-lipid sandwich does not allow the hydrophilic phosphate heads to be in contact with water, nor does it allow the non-polar hydrophobic amino acids on the outside of the membrane proteins to be kept away from water. Consideration of how lipids behaved in water, forming a bilayer because it is the most stable arrangement, was used to refine the model. Interpretation of the electron micrograph evidence changed to support the new model of the membrane structure. The phosphate heads are more electron dense and show up as the darker edges to the membrane, with the tails forming the lighter inner part of the sandwich.

Experiments showed that there were two types of proteins – those that could be dissociated from the membrane quite easily by increasing the ionic strength of the solution, and those (the majority) that could only be removed from the membrane by more drastic action such as adding detergents. This evidence supported the fluid mosaic model where some peripheral proteins are loosely attached on the outside surface of the membrane whilst integral membrane proteins are fully embedded within the phospholipids, some even spanning both layers. Several integral proteins were investigated further and shown to have regions at their ends that had polar hydrophilic amino acids, with the middle portion being mainly composed of non-polar hydrophobic amino acids (Figure 2.16B).

amino acids with hydrophobic side chains

amino acids with hydrophilic side chains

Figure 2.16 A The protein-lipid sandwich model. **B** In the fluid mosaic model integral membrane proteins have polar and non-polar regions.

Additional evidence for integral membrane proteins came from freeze-fracture electron microscopy studies. Frozen membrane sections were fractured along the weak point between the lipid layers (Figure 2.17), and the inner fractured surface coated in a heavy metal. Scanning electron microscopy, which gives three-dimensional images, revealed a smooth mosaic-like surface (the lipid tails) interspersed by much larger particles (the integral proteins) (Figure 2.18).

Figure 2.17 Freeze-fracture of membranes reveals the integral transmembrane proteins.

Finally, several experiments were carried out using labelled molecules that only attach to other specific molecules. In one experiment plant proteins called lectins that bind to polysaccharides were labelled with ferritin (a protein with a core of ferric oxide) which is visible under the electron microscope. When mixed with membrane samples, the lectins only bound to the outer surface of the membrane, and never to the inner. This showed that membranes are asymmetric, i.e. the outside surface of the membrane is different to the inside, and once again did not support the protein sandwich model, which would be the same on both sides.

Figure 2.18 Scanning electron micrograph showing integral transmembrane proteins.

67

Another neat experiment involved fusing mouse cells with human cells. Before the cells were fused a specific membrane protein was labelled in each cell type. The mouse membrane protein was given a green fluorescent label, and the human membrane protein was given a red fluorescent label. A light microscope was used to follow where the green and red proteins moved. Immediately after fusing the cells, the coloured labels remained in their respective halves, but after 40 minutes at 37 °C there was complete intermixing of the proteins (Figure 2.19). The only way the proteins could have intermixed was to have diffused through the membranes, showing the components were indeed fluid.

Q2.14   When the mouse and human membrane protein experiment was carried out at 15 °C, mixing was much slower. How does this evidence further support the fluid mosaic model of membrane structure?

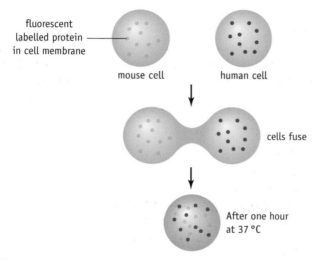

fluorescent labelled protein in cell membrane

mouse cell   human cell

cells fuse

After one hour at 37 °C

Figure 2.19 The movement of the fluorescent labelled proteins within the cell surface membranes supported the fluid mosaic model of membrane structure.

More unsaturated phospholipids – more fluid

The more phospholipids containing unsaturated fatty acids there are present in the membrane, the more fluid it is. The 'kinks' in the hydrocarbon tails of the unsaturated phospholipids prevent them from packing closely together, so more movement is possible. Cholesterol reduces the fluidity of the membrane by preventing movement of the phospholipids.

Q2.15   Can you suggest why the membrane is more fluid with unsaturated rather than saturated phospholipids making up the bilayer?

Many different types of protein are found within the membrane, each type having a specific function. Some function as enzymes. Others function as carrier and channel proteins involved in the transport of substances in and out of cells. Glycoproteins and glycolipids have important roles in cell-to-cell recognition and as receptors.

Activity

In **Activity 2.7** evaluate the evidence for different models that explain the structure and function of cell membranes. **A2.07S**. You can investigate the cell surface membrane practically in **Activity 2.8**. **A2.08S**

How do substances pass through cell membranes?

For a cell to function correctly it needs to be able to control transport across its surface membrane. Molecules and ions move across membranes by:

- diffusion
- osmosis
- active transport
- exocytosis
- endocytosis.

Diffusion

Diffusion is the net movement of molecules or ions from a region where they are at a higher concentration to a region of their lower concentration. Diffusion will continue until equilibrium, when the substance is evenly spread throughout the whole volume. Small uncharged particles diffuse across the cell membrane, passing between the lipid molecules as they move down the concentration gradient. Small molecules like oxygen and carbon dioxide can diffuse rapidly across the cell membrane. Carbon dioxide is polar but its small size still allows rapid diffusion.

Activity

Activity 2.9 lets you investigate different methods of transport. **A2.09S**

Facilitated diffusion

Hydrophilic molecules and ions that are larger than carbon dioxide cannot simply diffuse through the bilayer. They are insoluble in lipids, the hydrophobic tails of the phospholipids providing an impenetrable barrier to them. Instead they cross the membrane with the aid of proteins in a process called **facilitated diffusion**. They may diffuse through water-filled pores within **channel proteins** that span the membrane (Figure 2.15). There are different channel proteins for transporting different molecules.

Each type of channel protein has a specific shape that permits the passage of only one particular type of ion or molecule. Some channels can be opened or closed depending on the presence or absence of a signal, which could be a specific molecule, like a hormone, or a change in potential difference (voltage) across the membrane. These channels are called gated channels.

high concentration

low concentration

Some proteins that play a role in facilitated diffusion are not just simple channels but are **carrier proteins**. The ion or molecule binds onto a specific site on the protein. The protein changes shape (Figure 2.20) and as a result the ion or molecule crosses the membrane. The movement can occur in either direction, with the net movement being dependent on the concentration difference across the membrane. Molecules move from high to low concentration due to more frequent binding to carrier proteins on the side of the membrane where the concentration is higher.

Diffusion, whether facilitated or not, is sometimes called **passive transport**. 'Passive' here refers to the fact that no metabolic energy is needed for the transport.

Figure 2.20 The carrier protein changes shape, facilitating diffusion.

Osmosis

Osmosis is the net movement of water molecules from a solution with a lower concentration of solute to a solution with a higher concentration of solute through a partially permeable membrane. Osmosis can be summarised as follows:

Look at the left and right parts of Figure 2.21. In each case decide the direction in which the solvent (i.e. water) molecules will have a net (overall) tendency to move. Only the right half of Figure 2.21 has a partially permeable membrane so only here can osmosis take place. Osmosis is due to the random movement of water molecules across the membrane and is a particular type of diffusion. If solute molecules are present, water molecules form hydrogen bonds with them, and this reduces the movement of these water molecules. If more solute is present there are fewer water molecules free to collide with and move across the membrane.

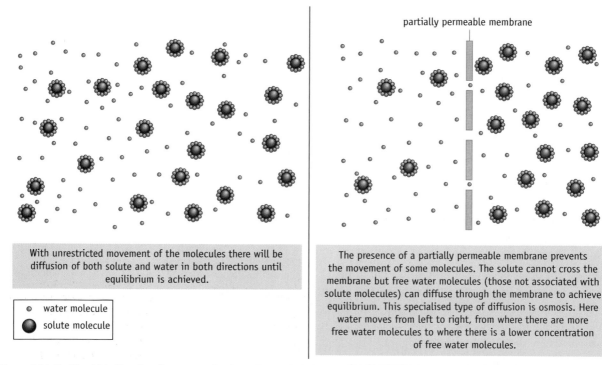

With unrestricted movement of the molecules there will be diffusion of both solute and water in both directions until equilibrium is achieved.

The presence of a partially permeable membrane prevents the movement of some molecules. The solute cannot cross the membrane but free water molecules (those not associated with solute molecules) can diffuse through the membrane to achieve equilibrium. This specialised type of diffusion is osmosis. Here water moves from left to right, from where there are more free water molecules to where there is a lower concentration of free water molecules.

○ water molecule
● solute molecule

Figure 2.21 Decide which direction the water molecules will move in before reading the shaded boxes to check if you were correct.

Active transport

If substances need to be moved across a membrane against a concentration gradient (from low concentration to high concentration) then energy is required. As with facilitated diffusion, specific carrier proteins are also needed. The energy comes from respiration and is supplied by the energy transfer molecule **ATP**. The substance to be transported binds to the carrier protein. Energy from ATP changes the shape of the carrier protein, causing the substance to be released on the other side of the membrane (Figure 2.22).

Active transport or pumping of substances across membranes occurs in every cell. Examples appear throughout the course, including transport of ions across epithelial cells (later in Topic 2), plant cell roots (Topic 4), muscle cells (Topic 6), and nerve cells (Topic 8). It also occurs between compartments within a cell, for example between the mitochondria and cytoplasm (Topics 3 and 7).

Figure 2.22 In active transport energy is required to change the shape of the carrier protein (dotted line) and move the substance across the membrane against a concentration gradient.

Exocytosis and endocytosis

Sometimes very large molecules or particles need to be transported across cell surface membranes. This is achieved by **exocytosis** and **endocytosis**, which rely on the fluid nature of the membrane (Figure 2.23). Exocytosis is the release of substances, usually proteins or polysaccharides, from the cell as vesicles (small membrane-bound sacs) fuse with the cell membrane. For example, insulin (the hormone produced by

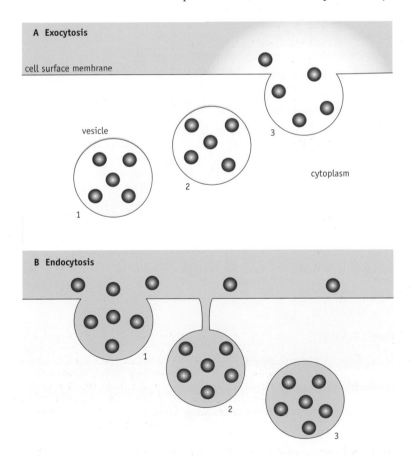

Figure 2.23 In exocytosis, vesicles fuse with the cell surface membrane, releasing their contents from the cell. In endocytosis, substances are brought inside a cell within vesicles formed from the cell surface membrane.

certain cells in the pancreas) is released into the blood by exocytosis. Neurotransmitter substances are also released in this way from nerve endings (Topic 8). Endocytosis is the reverse process: substances are taken into a cell by the creation of a vesicle. Part of the cell membrane engulfs the solid or liquid material to be transported. In some cases the substance to be absorbed attaches to a receptor in the membrane and is then absorbed by endocytosis. This is how cholesterol is taken up into cells. White blood cells ingest bacteria and other foreign particles by endocytosis (Topic 6).

Table 2.1 provides a summary of the different methods of transport across cell membranes.

Diffusion	high to low concentration until equilibrium reached
	hydrophobic (lipid-soluble) or small uncharged molecules
	through phospholipid bilayer
	passive, no energy required
Facilitated diffusion	high to low concentration until equilibrium reached
	hydrophilic molecules or ions
	through channel proteins or via carrier proteins that change shape
	passive, no energy required
Osmosis	a type of diffusion involving movement of free water molecules
	high to low concentration of free water molecules until equilibrium reached
	through phospholipid bilayer
	passive, no energy required
Active transport	against a concentration gradient, low to high concentration
	through carrier proteins that change shape
	requires energy
Exocytosis	used for bulk transport of substances out of the cell
	vesicles fuse with the cell surface membrane, releasing their contents
Endocytosis	used for bulk transport of substances into the cell
	vesicles are created from the cell surface membrane, bringing their contents into the cell

Table 2.1 Summary of methods of transport across cell membranes.

Q2.16 For each example below, suggest the type of transport most likely to be involved.

a movement of oxygen across the wall of an alveolus

b absorption of phosphate ions into root hair cells

c pumping of calcium ions into storage vesicles inside muscle cells

d release of glucose from liver cells into the bloodstream

e removal of the sodium ions that diffuse into a nerve cell, thus maintaining a low concentration within the nerve axon

f reabsorption of water molecules from the kidney tubule.

What happens in the membranes of the cells lining the airways?

The cells that line the airways produce mucus. In people who do not have cystic fibrosis, the amount of water in the mucus is continuously regulated to maintain a constant viscosity ('stickiness') of the mucus. It must be runny enough to be moved by the beating cilia but not so runny that the fluid floods the airway. This regulation of the water content of the mucus is achieved by the transport of sodium ions and chloride ions across the epithelial cells. Water then follows the ions because of osmosis.

Studies comparing the ionic composition of the airway surface liquid in CF and normal mice, and in CF and healthy humans, have provided evidence to support the theory that explains what is happening in the lungs of people with CF.

Regulating water in the mucus in unaffected lungs

Excess water in the mucus

If the mucus layer contains too much water this is detected by the membranes of cells lining the airways (epithelial cells). Carrier proteins in the basal membranes of the epithelial cells actively pump sodium ions out of the cells (Figure 2.24A). The concentration of sodium ions (Na^+) in the cell falls, setting up a concentration gradient across the apical membrane (the membrane facing the airway). Sodium ions diffuse down this concentration gradient. The ions pass into the cell by facilitated diffusion through sodium channels in the apical membrane.

The raised concentration of Na^+ in the tissue fluid on the basal membrane side of the epithelial cells creates a potential difference between this tissue fluid and the mucus on the apical membrane side. The tissue fluid now contains more positively charged ions than does the mucus. This creates an electrical gradient between the tissue fluid and the mucus. This electrical gradient causes negatively charged chloride ions to diffuse out of the mucus into the tissue fluid via the gaps between neighbouring epithelial cells.

The elevated Na^+ and Cl^- concentrations in the tissue fluid draw water out of the cell by osmosis across the basal membrane into the tissue fluid. This water loss increases the overall solute concentration within the cell. Since the solute concentration is now higher within the cell than in the mucus, water is drawn out of the mucus by osmosis across the apical membrane and into the cell.

With excess water

1 Na⁺ is actively pumped across the basal membrane.

2 Na⁺ diffuses through sodium channels in the apical membrane.

3 Cl⁻ diffuses down electrical gradient.

4 Water is drawn out of cells by osmosis due to the high salt concentration in the tissue fluid.

5 Water is drawn out of the mucus by osmosis.

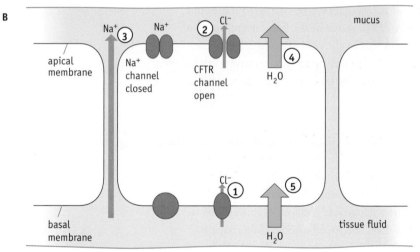

With too little water

1 Cl⁻ is pumped into the cell across the basal membrane.

2 Cl⁻ diffuses through the open CFTR channels.

3 Na⁺ diffuses down the electrical gradient into the mucus.

4 Elevated salt concentration in the mucus draws water out of the cell by osmosis.

5 Water is drawn into the cell by osmosis.

With CF

1 CFTR channel is absent or not functional.

2 Na⁺ channel is permanently open.

3 Water is continually removed from mucus by osmosis.

Figure 2.24 The role of the CFTR channel in keeping the mucus runny.

Having too much water in the mucus is a normal state of affairs. This is because the cilia are continuously moving mucus along the airways. Movement of mucus from numerous smaller bronchiole branches into fewer larger bronchioles means that water must be removed to reduce the volume of mucus so as to avoid the larger airways flooding with fluid.

Too little water in the mucus

When there is too little water in the mucus, chloride ions are transported across the basal membrane into the epithelial cell (Figure 2.24B). This creates a concentration gradient across the apical membrane, with the concentration of chloride ions being higher inside the cell than out. At the same time, the cystic fibrosis transmembrane regulatory (CFTR) protein channels (a type of gated channel) open. Chloride ions diffuse out of the cell through the CFTR channels down this concentration gradient into the mucus. When open, the CFTR channels block (close) the sodium ion channels in the apical membrane. The mechanism for this is currently unknown. The build-up of negatively charged chloride ions in the mucus creates an electrical gradient between the mucus and the tissue fluid. Sodium ions diffuse out of the tissue fluid and move down this electrical gradient, passing between the cells into the mucus. The movement of the sodium and chloride ions into the mucus draws water out of the cells by osmosis until the solutions on either side of the membrane have the same concentration of free water molecules (until they are isotonic). This movement of water prevents the mucus that lines the airways from becoming too viscous (sticky).

Why CF lungs cannot regulate the water in mucus

In a person who has CF, the CFTR protein may be missing, or if present it does not function correctly (Figure 2.24C). When there is too little water in the mucus, Cl⁻ cannot be secreted across the apical membrane, and there is no blockage of the epithelial sodium ion channels. Since the Na⁺ channels are always open, there is continual Na⁺ absorption by the epithelial cells. The raised levels of Na⁺ draw chloride ions and water out of the mucus into the cells. This makes the mucus more viscous, which makes it harder for the beating cilia to move it, so the mucus is not effectively cleared up and out of the lungs. Sticky mucus builds up in the airways. This mucus frequently becomes infected with bacteria, causing a downward spiral of airway inflammation and damage.

Activity

Work through the interactive tutorial in **Activity 2.10** to investigate the effects of a functioning and a non-functioning CFTR protein channel on salt and water secretion in the airways. **A2.10S**

2.3 How does cystic fibrosis affect other body systems?

The effect of CF on the digestive system

Cystic fibrosis sufferers have difficulty maintaining body mass because of problems with the digestion and absorption of nutrients. They also have high basal metabolic rates. They generally have poor appetites but still have to eat than most people, including high-energy food to make sure they obtain sufficient nutrients and energy. They require 120–140% of the recommended daily energy intake. People with CF may also take food supplements that contain digestive enzymes. The aim of these supplements is to help break down large food molecules.

Most of the chemical breakdown of food molecules and the subsequent absorption of the soluble products into the bloodstream occurs in the small intestine. Glands secrete digestive enzymes into the lumen of the gut, where they act as catalysts to speed up the extracellular breakdown of food molecules. A wide range of enzymes are produced by exocrine glands (Figure 2.25) outside the gut, e.g. salivary glands, the liver and the pancreas. Enzymes are also built into the membranes of the gut wall.

Groups of pancreatic cells produce enzymes that help in the breakdown of proteins, carbohydrates and lipids. These digestive enzymes are delivered to the gut in pancreatic juice released through the pancreatic duct (Figure 2.26).

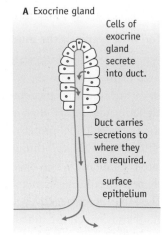

A Exocrine gland

Cells of exocrine gland secrete into duct.

Duct carries secretions to where they are required.

surface epithelium

B Endocrine gland

Cells of endocrine gland secrete directly into the blood.

blood capillary

Figure 2.25 There are two types of gland. These are: **A** exocrine glands with ducts and **B** endocrine glands which are ductless. Endocrine glands release hormones.

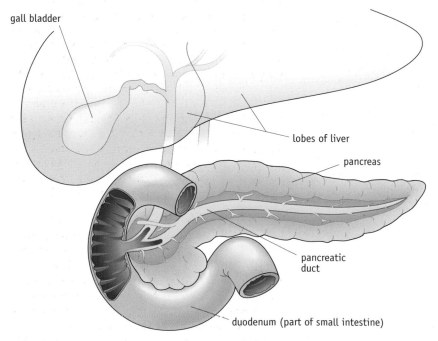

gall bladder

lobes of liver

pancreas

pancreatic duct

duodenum (part of small intestine)

Figure 2.26 Thick mucus can easily block the pancreatic duct in a CF patient.

In a person with CF, the pancreatic duct becomes blocked by sticky mucus, impairing the release of digestive enzymes. The lower concentration of enzymes within the small intestine reduces the rate of digestion. Food is not fully digested, so not all the nutrients can be absorbed. A higher proportion of partially digested and undigested food means energy is lost in the faeces. This is called malabsorption syndrome.

An additional complication occurs when the pancreatic enzymes become trapped behind the mucus blocking the pancreatic duct. These enzymes damage the pancreas itself. The cysts of hard, damaged or fibrosed tissue within the pancreas give the CF condition its name. Another complication occurs if damage occurs to cells within the pancreas that produce the hormone insulin, which is involved in the control of blood sugar levels. A form of diabetes can be the result.

Key biological principle: Enzyme function depends on protein three-dimensional structure

Enzymes are globular proteins that act as biological **catalysts**. They speed up chemical reactions that would otherwise occur very slowly at the temperature within cells. The precise three-dimensional shape adopted by an enzyme includes a depression on the surface of the molecule called the **active site**. This site may be a relatively small part of the large protein molecule, as you can see in Figure 2.27. Only a few amino acids may be directly involved in the active site, with the remainder maintaining the three-dimensional shape of the protein molecule.

Lock-and-key theory

Either a single molecule with a complementary shape, or more than one molecule that together have a complementary shape, can fit into the active site (Figure 2.28, overleaf). These **substrate** molecule(s) form temporary bonds with the amino acids of the active site to produce an **enzyme-substrate complex**. The enzyme holds the substrate molecule(s) in such a way that they react more easily. When the reaction has taken place the products are released, leaving the enzyme unchanged. The substrate is often likened to a 'key' which fits into the enzyme's 'lock', so this is known as the **lock-and-key theory** of enzyme action. Each enzyme will only catalyse one specific reaction because only one shape of substrate will fit into its precisely-shaped active site.

Induced fit theory

It has been found that the active site is often flexible. When the substrate (or substrates) enters the active site, the enzyme molecule changes shape slightly, fitting more closely around the substrate (Figure 2.27). It is like a person putting on a wetsuit; the wetsuit shape changes to fit the body but returns to its original shape when taken off. This is known as the **induced fit theory** of enzyme action. Only a specific substrate will induce the change in shape of an enzyme's active site.

Activation energy

To convert substrate(s) into product(s), bonds must change both within and between molecules. Breaking chemical bonds requires energy whilst energy is released when bonds form. The energy needed to break bonds and start the reaction is known as the **activation energy**. Without an enzyme, heating a substrate would provide this energy. (Think about starting a bonfire – a reaction between the molecules in wood and oxygen. You must first provide some energy to start the fire.) The heat energy agitates atoms within the molecules, the molecules become unstable and the reaction can then proceed. In cells, enzymes reduce the amount of energy needed to bring about a reaction; this allows reactions to occur without raising the temperature of the cell.

How do enzymes reduce the activation energy?

The specific shape of the enzyme's active site and of its complementary substrate(s) is such that electrically charged groups on their surfaces interact (Figure 2.28, overleaf). The attraction of oppositely charged groups may distort the shape of the substrate(s) and assist in the breaking of bonds or in formation of new bonds. In some cases, the active site may contain amino acids with acidic side chains; the acidic environment created within the active site may provide conditions favourable for the reaction.

Enzymes are present in all organisms and catalyse a huge range of reactions. They speed up reactions by at least a million times and most biological reactions

Figure 2.27 The active site is only a small part of the enzyme molecule. Here, the active site of the enzyme hexokinase changes shape as the glucose (yellow) enters.

CONTINUED ▶

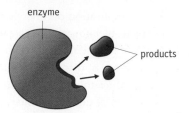

enzyme molecule

active site – cleft in
enzyme molecule

substrate
molecule

enzyme

enzyme

products

1 Random movement
causes the enzyme and
substrate to collide,
and the substrate
enters the active site.

2 Enzyme-substrate complex
forms. Charged groups
attract, distorting the
substrate and aiding bond
breakage or formation.

3 Products are released from
the active site leaving the
enzyme unchanged and
ready to accept another
substrate molecule.

Figure 2.28 Charged side groups interact and assist in breaking or forming bonds.

wouldn't happen at all without enzymes. The metabolism of an organism is the sum of all the enzyme-catalysed reactions occurring within it. Some metabolic reactions are extremely simple, like the hydration of carbon dioxide, which is catalysed by carbonic anhydrase:

Carbonic anhydrase

$$CO_2 + H_2O \rightleftharpoons H_2CO_3$$

This reaction allows carbon dioxide to be transported in the blood from respiring tissues and then out into the alveoli. Enzymes catalyse many different types of reactions, some occurring inside cells (intracellular reactions) and others in tissue fluid, blood or other aqueous solutions in the organism (extracellular reactions). Sometimes large substrate molecules will be broken down into smaller units in catabolic (breaking down) reactions. The hydrolysis of starch to maltose by salivary amylase is an extracellular catabolic reaction. Other enzymes catalyse anabolic (building up) reactions,

like the fatty acid synthetase enzyme complex that, unsurprisingly, synthesises fatty acids within cells. Sometimes substrate molecules are simply altered, not added to or broken down – isomerase enzymes do this.

As we have seen, enzymes:

- are globular proteins
- have an active site that allows binding with a specific substrate
- catalyse (speed up) reactions
- reduce the activation energy required for a chemical reaction to take place
- do not alter the end-product or nature of a reaction
- remain unchanged at the end of a reaction, able to bind with another substrate molecule.

Finding rates of enzyme-controlled reactions

The rate of reaction is measured by determining the quantity of substrate used or the quantity of product formed in a given time. For example, when the enzyme catalase is used to break down hydrogen peroxide, H_2O_2, to water and oxygen, the rate of reaction can be found by measuring the volume of oxygen given off in a known time.

If we mix a fixed quantity of enzyme and substrate, at first the reaction will proceed quickly, as shown in Figure 2.29. However, as the substrate is used up, there are fewer substrate molecules to bind with the enzyme and the reaction slows down and eventually stops (no further increase in the product occurs), as seen in Figure 2.30B. The slope of the rapid phase of the reaction is known as the initial rate of reaction and is frequently used when comparing rates of enzyme-controlled reactions.

Checkpoint ✓

2.3 Write a definition for each of the following key enzyme terms:

- biological catalyst
- activation energy
- active site
- enzyme-substrate complex.
- product
- lock-and-key theory
- induced fit theory

2.4 Explain how the three-dimensional structures of proteins enable enzymes to perform their functions as biological catalysts.

CONTINUED ▶

Figure 2.29 The quantity of product is measured over time to determine the progress of an enzyme-catalysed reaction. Here the volume of oxygen produced is measured when catalase is added to hydrogen peroxide.

How do enzyme and substrate concentrations affect the rate of reaction?

Figure 2.30A shows the effect of enzyme concentration on the initial rate of reaction. The initial rate of reaction is directly proportional to the enzyme concentration because the more enzyme that is present, the greater the number of active sites that are available to form enzyme-substrate complexes. The increase in rate will continue in this linear fashion assuming that there is an excess of substrate. In Figure 2.30B you can see that at high substrate concentrations it is the enzyme concentration that limits the rate of reaction. Every active site is occupied and substrate molecules cannot enter an active site until one becomes free again.

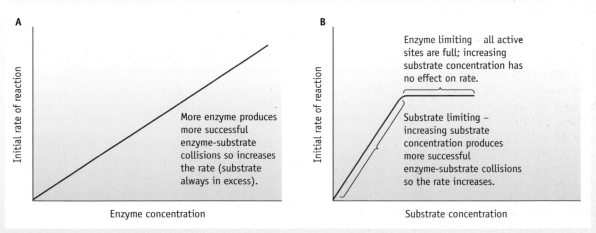

Figure 2.30 The effect of A enzyme concentration B substrate concentration on the rate of an enzyme-controlled reaction.

Activity

In **Activity 2.11** you can investigate the effect of enzyme concentration on enzyme activity. **A2.11S**

The effect of CF on the reproductive system

We have seen how sticky mucus produced by a defective CFTR protein can lead to major complications in the lungs and pancreas. In the reproductive system it can also cause severe problems. Females have a reduced chance of becoming pregnant because a mucus plug develops in the cervix (Figure 2.31A). This stops sperm from reaching the egg. Males with cystic fibrosis commonly lack the vas deferens (sperm duct) on both sides, which means sperm cannot leave the testes (Figure 2.31B). Where the vas deferens is present it can become partially blocked by a thick sticky mucus layer. This means fewer sperm are present in each ejaculate.

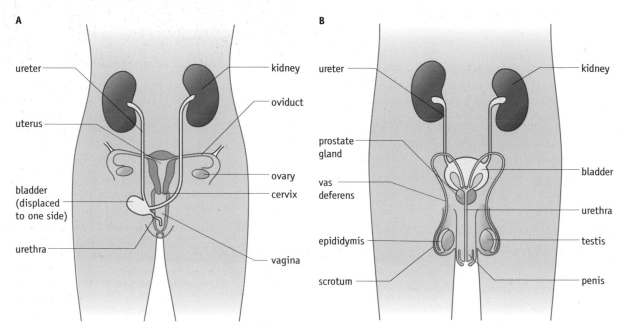

Figure 2.31 A The female reproductive system. In CF a mucus plug develops in the cervix. **B** The male reproductive system. In CF the vas deferens may be absent or plugged with mucus.

Q2.17 Unusually salty sweat is often one of the first signs that a baby may have CF. Why might the sweat of a person with CF be more salty than normal? (Hint: in sweat ducts the CFTR protein works in the opposite direction to the way it works in mucus-producing cells.)

Checkpoint

2.5 For each of **a** to **c**, write five bullet point statements to describe the effect CF has on:
a the gas exchange system **b** the digestive system **c** the reproductive system.

2.4 How is the CFTR protein made?

Cystic fibrosis is caused by a **mutation** in the DNA that carries the instructions for making the CFTR protein. In order to know how DNA works (and what has gone wrong in cystic fibrosis), you need to know what a gene is and how the codes it contains are used to make proteins.

The structure of DNA

In 1953 James Watson and Francis Crick (Figure 2.32) proposed a model for the structure of DNA, using evidence from the X-ray diffraction patterns of DNA produced by Rosalind Franklin and Maurice Wilkins. Their model was correct and their discovery has revolutionised biology.

DNA is found in every cell nucleus. It contains the genetic code which dictates all the inherited characteristics of an organism. It does this by controlling the manufacture of proteins. Your proteins, coded for by your DNA, are what make you unique; they are what make you a human being and not an oak tree or a chimpanzee. They play a vital role in giving you the unique characteristics that mean you are not the same as the person sitting next to you (you may be rather glad about that!).

Gene and genome

A **gene** is a sequence of bases on a DNA molecule coding for a sequence of amino acids in a polypeptide chain. Each chromosome found in the cell nucleus contains a large amount of DNA and carries numerous genes. The genes make up only a fraction of the total length of DNA in the chromosomes; the job of the remainder of the DNA is not fully known. Together, all the genes in an individual (or species) are known as the **genome**.

DNA is a chain of nucleotides

DNA is one type of nucleic acid, called **deoxyribonucleic acid**. It is a long chain molecule made of many units called nucleotides or mononucleotides.

Figure 2.32 James Watson (left) and Francis Crick (right) worked out the structure of DNA in 1953. 'The seed of life itself. Peel the chains apart, each chain reproduces the other, one becomes two, two become one. Generation on generation, all the way from Adam and Eve to you and me. It never dies. One simple shape. The womb of humanity. Endlessly, effortlessly fertile, dividing, reforming... It's the closest we'll ever get to immortality.' (Source: Tim Piggott-Smith, who played Francis Crick in the film *Life Story*, talking about the structure of DNA.)

> **Extension**
>
> You can read about the controversy surrounding the work of Rosalind Franklin and DNA in **Extension 2.1. X2.01S**

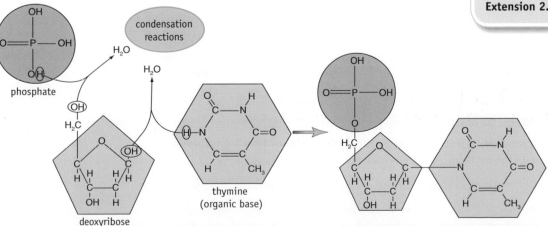

Figure 2.33 A deoxyribose sugar, a phosphate and a base join to form a mononucleotide (often known as a nucleotide).

Q2.18 DNA is called a polynucleotide. Explain why.

A mononucleotide contains three molecules linked together by condensation reactions: **deoxyribose** (a 5-carbon sugar), a **phosphate group** and an **organic base** containing nitrogen. Look at Figure 2.33 (previous page) to see how these three are arranged in a mononucleotide.

Mononucleotides link together by condensation reactions between the sugar of one nucleotide and the phosphate of the next one, producing a long chain of nucleotides, a polynucleotide. The nitrogen-containing base is the only part of the nucleotide that is variable. There are four bases – **adenine**, **cytosine**, **guanine** and **thymine**. These bases are often represented by their initial letters: **A**, **C**, **G** and **T** respectively.

In a DNA molecule there are two long strands of nucleotides twisted around each other to form a double helix (Figure 2.34), rather like a spiral staircase. The sugars and phosphates form the 'backbone' of the molecule and are on the outside. The bases point inwards horizontally, rather like the rungs of a ladder. The two strands, which run in opposite directions, are known as antiparallel strands and are held together by hydrogen bonds between pairs of bases. The DNA in each human cell contains some 3000 million of these base pairs.

Support

To remind yourself about hydrogen bonds, visit the Biochemistry support on the website.

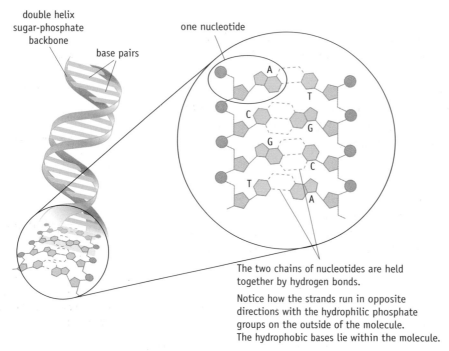

double helix
sugar-phosphate backbone
base pairs
one nucleotide

The two chains of nucleotides are held together by hydrogen bonds.

Notice how the strands run in opposite directions with the hydrophilic phosphate groups on the outside of the molecule. The hydrophobic bases lie within the molecule.

Figure 2.34 The structure of DNA. Remember that any picture you see of DNA is a simplification, a model based on evidence from techniques such as X-ray diffraction. A whole DNA molecule is much longer than this and usually contains millions of base pairs.

Why do the bases pair up?

If you look at Figure 2.34 you should notice that the bases only pair in a certain way: adenine only pairs with thymine, and cytosine only pairs with guanine.

Q2.19 Look at Figure 2.34 and suggest why the bases might only form these pairs.

The key to this pairing is in the structure of the bases and the bonding between them. Bases A and G both have a two-ring structure, whereas C and T have only one ring. The bases pair so that there are effectively three rings forming each 'rung' of the DNA molecule, making the molecule a uniform width along its whole length. The shape and chemical structure of the bases dictates how many hydrogen bonds each one can form and this determines the pairing of A with T (two hydrogen bonds) and C with G (three hydrogen bonds). This deceptively simple fact is the clue to how DNA works. The bases A and T and the bases C and G are referred to as complementary base pairs.

Q2.20   The sequence of bases on part of one strand of a DNA molecule is:

A T C C C T G A G G T C A G T

What would be the sequence of bases on the corresponding part of the other strand?

How does DNA code for proteins?

The CF gene is on chromosome 7. It is a long gene, made up of about 230 kbp (230 000 base pairs), and it instructs the cell to make the CFTR protein that forms the transmembrane chloride channel. But what is the genetic code? The sequence of bases in the DNA tells the cell which amino acids to link together to make the CFTR protein. Each gene is a sequence of bases on a DNA molecule coding for a sequence of amino acids in a polypeptide chain.

The triplet code

In the genetic code, one base does not simply code for one amino acid. There are only four bases, so if this were the case proteins could contain only four different amino acids, instead of the 20 amino acids found commonly in proteins. The code carried by the DNA is a three-base or **triplet code** (Figure 2.35). Each adjacent group of three bases codes for an amino acid. Several triplets code for the same amino acid; others are start signals or stop signals (called chain terminators). This amazingly simple but fundamental coding system is found in all organisms.

From DNA to proteins

The DNA in our chromosomes carries the genetic information from one generation to the next. It carries the codes that help determine the structure and function of cells by telling the cells which proteins to make. But DNA is in the nucleus, and proteins are made in the cytoplasm. DNA cannot pass through the membranes surrounding the nucleus into the cytoplasm. So how do the instructions get from nucleus to cytoplasm?

A 'copy' of the DNA is made. It is rather like making a photocopy of some precious original plans so that everyone on the factory floor can work from them, leaving the originals safe at head office. This 'copy' is not made from DNA but from another type of nucleic acid called **ribonucleic acid**, **RNA**. This RNA can leave the nucleus, carrying the information to the cytoplasm where it is used in the manufacture of proteins.

What is the difference between DNA and RNA?

An RNA molecule has a *single* strand made of a string of RNA nucleotides. These are very similar in structure to DNA nucleotides except that they contain **ribose** sugar and not deoxyribose (Figure 2.36). Another difference is that in RNA nucleotides, the base **uracil** (**U**) replaces thymine, so RNA *never* contains thymine.

Activity

Try making a 3D model of DNA in **Activity 2.12. A2.12S**

Extract DNA from onions in **Activity 2.13. A2.13S**

Extension

In **Extension 2.2** work out the code used on the DNA. **X2.02S**

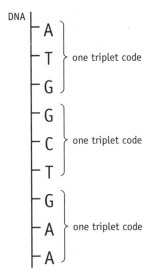

Figure 2.35 Each DNA triplet of three bases codes for one amino acid. Several triplets code for the same amino acids.

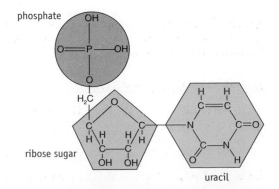

Figure 2.36 Compare this RNA nucleotide with the DNA nucleotide in Figure 2.33 and notice the differences.

There are three types of RNA involved in protein synthesis: **messenger RNA (mRNA)**, **transfer RNA (tRNA)** and **ribosomal RNA (rRNA)** as outlined below.

Protein synthesis

Protein synthesis has two stages; the first occurs in the nucleus and the second in the cytoplasm (Figure 2.37).

1 Transcription

Transcription takes place in the nucleus. The DNA double helix unwinds and hydrogen bonds between the bases break, allowing the two strands to partly separate. The sequence on one of the strands, the **template strand**, is used in the production of a messenger RNA molecule. This mRNA is built from free RNA nucleotides which line up alongside the DNA template strand. Because of complementary base pairing the order of bases on the DNA exactly determines the order of the bases on the RNA. In other words, every triplet code on DNA gives rise to a complementary **codon** on messenger RNA (Figure 2.39).

The template strand is also known as the antisense strand because once transcribed it makes an mRNA molecule with the same base sequence as the DNA coding strand. The coding strand is known as the sense strand. Be aware that many biology text books are incorrect in their use of the terms sense strand and antisense strand.

This process of synthesising involves a number of enzymes such as RNA polymerase. The completed messenger RNA molecule now leaves the nucleus through a pore in the **nuclear envelope** (the two membranes that surround the nucleus) and enters the cytoplasm. This is where the second stage takes place.

2 Translation

Translation takes place on **ribosomes**. Ribosomes are small organelles (structures within the cell) made of ribosomal RNA and protein. Ribosomes are found free in the cytoplasm or attached to **endoplasmic reticulum**, a system of flattened, membrane-bound sacs. A transfer RNA molecule carrying an amino acid molecule has three bases called an **anticodon** (Figure 2.38) and these pair with complementary bases on the mRNA codon. Then the amino acids that the tRNA carry join by means of peptide bonds (Figure 2.39).

Activity

The computer simulation in **Activity 2.14** looks at nucleic acid structure and the sequence of events in protein synthesis. **A2.14S**

Figure 2.37 In transcription, the DNA code is copied onto mRNA. In translation, it is used to join the correct sequence of amino acids and build a new protein.

amino acid site
hydrogen bonds between paired bases

anticodon

This can be simplified to:

amino acid binding site   amino acid binding site

anticodon   anticodon

Figure 2.38 The tRNA molecule carries an amino acid. Its anticodon base pairs with the codon on the mRNA molecule.

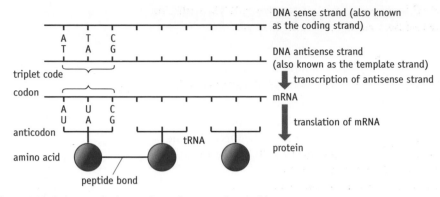

Figure 2.39 Codons, anticodons and complementary base pairing.

A protein molecule is usually made up of between 50 and 2000 amino acids. The sequence of the amino acids, its primary structure, determines the three-dimensional structure and properties of the protein. A slight variation in the order, for example changing just one of the amino acids in the chain, may substantially alter the protein's structure and properties.

Q2.21 On which type of RNA would you find **a** a codon **b** an anticodon?

Q2.22 The sequence of bases AGT form a triplet code on the sense strand. What is:

a its triplet code on the antisense strand

b its codon

c its anticodon

Figure 2.40 False-colour transmission electron micrograph showing translation, new proteins (blue) are formed from the mRNA (black). As the protein molecule is formed, it folds up into the three-dimensional shape determined by its primary structure – the sequence of amino acids (see page 62).

Q2.23 What would be the sequence of bases on a length of messenger RNA built using the following DNA strand as a template?

T A C A T G G A T T C C G A T

Q2.24 How many tRNA molecules would be involved in the synthesis of the protein coded for by this section of DNA?

Q2.25 What are the anticodons, assuming you read the section from left to right?

2.5 What goes wrong with DNA?

A mistake in translation can produce mRNA with one or more incorrect codons. This could result in the production of a faulty protein or no protein at all. But because the fault is in the mRNA, it would only affect the proteins produced from this one mRNA strand in this one cell, on this one occasion. It would not produce the problems seen in every epithelial cell of cystic fibrosis sufferers. It is errors in the DNA that are responsible for inherited genetic conditions. These mistakes arise when DNA copies itself during the process of DNA replication.

DNA replication

When a cell divides, an exact copy of the DNA must be produced so that each of the daughter cells receives a copy. This process of copying the DNA is called **replication**.

The DNA double helix unwinds from one end and the two strands split apart as the hydrogen bonds between the bases break (Figure 2.41). Free DNA nucleotides line up alongside each DNA strand and hydrogen bonds form between the complementary bases. The enzyme DNA polymerase links the adjacent nucleotides to form a complementary strand. In this way each strand of DNA acts as a template on which a new strand is built and, overall, two complete DNA molecules are formed. These are identical to each other and to the original DNA molecule. Check this by comparing them in Figure 2.41. Each of the two DNA molecules now contains one 'old' strand and one 'new' strand. This process is therefore known as **semi-conservative replication**.

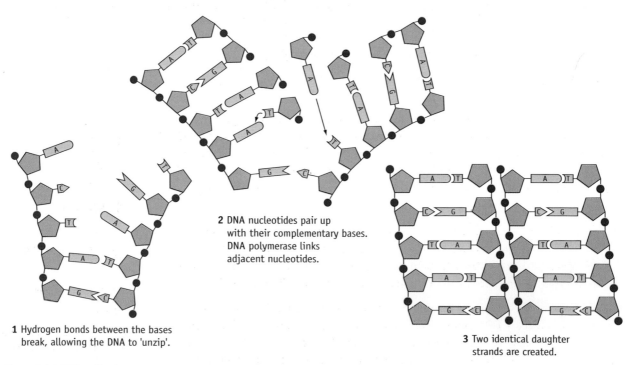

1 Hydrogen bonds between the bases break, allowing the DNA to 'unzip'.

2 DNA nucleotides pair up with their complementary bases. DNA polymerase links adjacent nucleotides.

3 Two identical daughter strands are created.

Figure 2.41 DNA replication.

How do we know that DNA replication is semi-conservative?

In their letter to the science journal 'Nature' in 1953, proposing the double helical model of DNA, Watson and Crick wrote:

> *'It has not escaped our notice that the specific pairing we have postulated immediately suggests a possible copying mechanism for the genetic material.'*
>
> (*Source:* Watson, J.D. and Crick, F.H.C., 'Molecular structure of nucleic acid. A structure for deoxyribose nucleic acid', *Nature* 171 pp 737–738 (1953).)

However, it took four more years and a series of elegant experiments by the American geneticists Matthew Meselson and Franklin Stahl to confirm experimentally what Watson and Crick suspected about how this copying mechanism happens.

There are three possible ways DNA *could* replicate – in a fragmentary way, semi-conservatively or conservatively (Figure 2.42).

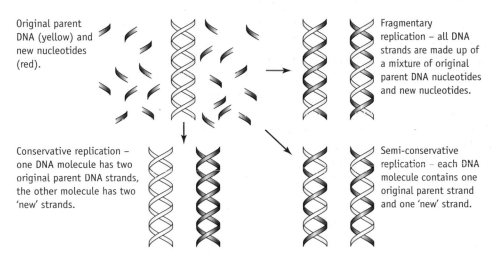

Original parent DNA (yellow) and new nucleotides (red).

Fragmentary replication – all DNA strands are made up of a mixture of original parent DNA nucleotides and new nucleotides.

Conservative replication – one DNA molecule has two original parent DNA strands, the other molecule has two 'new' strands.

Semi-conservative replication – each DNA molecule contains one original parent strand and one 'new' strand.

Figure 2.42 Three possible models of DNA replication.

Figure 2.42 uses colour in the different DNA strands to follow their progress; Meselson and Stahl used heavy and light strands of DNA. They did this by using DNA from *Escherichia coli* bacteria that had been grown in a medium containing only the heavy isotope of nitrogen, ^{15}N. All the nucleotides in the bacteria at the start of their experiment contained heavy nitrogen, and this made the DNA denser than normal (yellow in Figure 2.42).

Meselson and Stahl then moved the bacteria into a medium containing only normal ^{14}N. This meant all the new nucleotides incorporated into the replicated DNA were 'light' but the original DNA nucleotides were heavy. They allowed the bacteria to divide and their DNA to replicate once. They then extracted and centrifuged the DNA.

If a test tube containing DNA dissolved in a special density-gradient solution is centrifuged, the heavy DNA (containing only ^{15}N) sinks to the bottom. Light DNA (containing only normal ^{14}N) collects in a band near the top, and DNA of medium density is in the middle (Figure 2.43). Medium density DNA must contain some heavy and some light nucleotides.

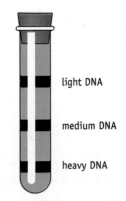

light DNA

medium DNA

heavy DNA

Figure 2.43 Position of DNA after density-gradient centrifugation.

Q2.26 Where would you expect the bands of DNA to be after one round of replication according to each of the theories described in Figure 2.42?

Meselson and Stahl's result was a single band of medium density DNA (Figure 2.44A). No heavy DNA existed any longer, so they were able to reject the conservative model of DNA replication. However, which of the other two models was correct, as both would yield medium DNA after one round of replication? The answer lay in letting the bacteria undergo two rounds of replication.

Q2.27 Using two different colours to represent the heavy and light bands, draw the DNA molecules after a second round of **a** fragmentary replication **b** semi-conservative replication.

Figure 2.44 A Position of DNA after one replication. **B** Position of DNA after replicating twice.

The DNA extracted and centrifuged after two rounds of replication gave two bands – one medium and one light (Figure 2.44B). The presence of both medium and light DNA bonds confirmed the semi-conservative model and ruled out the fragmentary model which would only produce one type of DNA containing a mixture of heavy and light nucleotides.

Q2.28 Where would the DNA be in the test tube if the fragmentary model was correct?

Activity

Look at the evidence that supports and refutes the different theories for replication in **Activity 2.15. A2.15S**

Mistakes in replication

So what has all this to do with cystic fibrosis and other genetic diseases? Sometimes DNA replication does not work perfectly. As the 'new' strand of DNA is being built, an incorrect base may slip into place. This is an example of a gene **mutation**.

Q2.29 Assuming no mutations, what would be the sequence of bases on the complementary strand created by replication of an 'old' strand with the sequence

C A G T C A G G C?

Q2.30 Identify the mutations that have occurred in replication of the same sequence in each of the following:

a G T C A G G C C G

b G A C A G T C C G

c G T C A T G C C G

d G T C G T C C G

e G T C A G G T C C G

Sometimes mutations occur in the DNA of an ovary or testis cell that is dividing to form an egg or sperm. Such a mutation may be passed on to future generations, present in every single cell produced from the fertilised egg.

Some mutations have no effect on the organism. Large amounts of the DNA found in a cell do not actually play a role in protein synthesis and therefore mutations that occur in these sections may have no effect. However, if a mutation occurs within a gene and a new base triplet is created that codes for a stop signal or a different amino acid, the protein formed may be faulty. This could cause a genetic disorder.

A mutation causes sickle cell anaemia

In the disease **sickle cell anaemia**, there is a mutation in the gene that codes for one of the polypeptide chains in haemoglobin, the pigment in red blood cells which carries oxygen around the body. The base adenine replaces thymine at one position along the chain. The mRNA produced from this DNA contains the triplet code GUA rather than GAA. As a result the protein produced contains the non-polar amino acid valine rather than polar glutamic acid at this point. This small change has a devastating effect on the functioning of the molecule. The haemoglobin is less soluble. When oxygen levels are low, the molecules form long fibres that stick together inside the red blood cell, distorting its shape. The half moon (sickle) shaped cells carry less oxygen and can block blood vessels (Figure 2.45).

Figure 2.45 Can you identify the sickle-shaped red blood cells, symptoms of sickle cell anaemia? Magnification ×1650.

Mutations and cystic fibrosis

The CF gene is a section of DNA on chromosome 7 carrying the code to make the CFTR protein. The CFTR protein contains 1480 amino acids and these are arranged into the 3D structure shown in Figure 2.46.

extracellular loops

six sections composed of helices span the lipid bilayer

intracellular loops

lipid bilayer

channel

ATP binding site 1. The most common CF mutation, DF508, occurs in this region of the protein.

The channel is only active when this R region is phosphorylated (i.e. has phosphate attached).

ATP binding site 2

Figure 2.46 The CFTR protein is a channel protein.

Cystic fibrosis is not as straightforward a genetic story as sickle cell anaemia. In the CF gene hundreds of different mutations have been identified that can give rise to cystic fibrosis. The mutations affect the CFTR protein in different ways. In some cases ATP is unable to bind and open the ion channel; in other cases the channel is open but changes in the protein structure lead to reduced movement of chloride ions through the channel. The most common mutation, known as the DF508 mutation, is the deletion of three nucleotides. This causes the loss of phenylalanine, the 508th amino acid in the CFTR protein, which is thought to result in misfolding of the protein. The mutations are passed from parent to offspring.

Do Claire and Nathan have one of these mutations? Will they pass it on and if they do, will the child have the disease?

Weblink

You can find out more about the mutations that cause cystic fibrosis by visiting the Oxford Gene Medicine Research Group website.

2.6 How is cystic fibrosis inherited?

Claire and Nathan know that cystic fibrosis is inherited but they need to understand *how* it is inherited. This will tell them the chances of their children inheriting the disease.

How genes are passed on

Genes and chromosome pairs

You know that a gene is a length of DNA that codes for a protein. Every cell (except the sex cells) contains *two* copies of each gene, one from each parent. For any particular gene the two copies are located in the same position, or **locus**, on each of two paired chromosomes. We have 23 pairs of chromosomes in our cells, and the chromosomes in each pair are called **homologous chromosomes**. Within each of these pairs, one chromosome comes from our mother and the other from our father.

Cystic fibrosis is caused by a gene mutation passed on from parents to their children. But consider the following three situations:

1 A couple has six children. The first five are healthy; the sixth has cystic fibrosis. Neither parent has the disease.

2 Another couple has two children; both children have cystic fibrosis. Again, neither parent has the disease.

3 A woman with cystic fibrosis is told that it is unlikely that her children will have the disease, but they will all be 'carriers' and could themselves have children with the illness.

So what is going on?

Genotypes, phenotypes and alleles

As we have seen, cystic fibrosis is caused by a mutation in the CF gene, the length of DNA that codes for the CFTR protein. The CF gene occurs in two alternative forms or **alleles**. First, there is the normal allele which codes for the functioning CFTR protein; this can be represented by the letter **F**. Secondly, there is the mutated allele which produces a non-functional protein; this can be represented by the letter **f**. Since every human has two copies of this gene in all their body cells there are three possible combinations that can occur, namely:

1 **FF** – A person with two identical copies of the normal allele does not have cystic fibrosis.

2 **ff** – A person with two copies of the mutated allele has cystic fibrosis.

3 **Ff** – A person with one normal allele and one mutated allele does not have cystic fibrosis but is a **carrier** and could have children who have the disease.

The alleles that a person has make up their **genotype**. Combinations 1 and 2, **FF** and **ff**, show a **homozygous** genotype for the CF gene – there are two identical copies of the allele. Combination 3 has two different alleles, **Ff**, and is **heterozygous** for the CF gene (Figure 2.47).

The characteristic caused by the genotype, i.e. its observable effect, is the **phenotype**. Table 2.2 summarises the cystic fibrosis genotypes and phenotypes.

homologous pair of chromosomes

gene locus – the position of a gene on a chromosome

centromere

alleles – alternative forms of the same gene

homozygous conditions

heterozygous condition

The dominant allele is written as an upper case (capital) letter.

Figure 2.47 Some basic genetic vocabulary.

Genotype	Phenotype
FF	normal
ff	cystic fibrosis
Ff	normal, but carrier

Table 2.2 The relationship between genotype and phenotype for the cystic fibrosis gene.

F is called the **dominant allele**. It affects the phenotypes of one of the homozygotes (a person who has the homozygous genotype, **FF**) and the heterozygote (a person with the heterozygous genotype, **Ff**). On the other hand, **f** is the **recessive allele**; it only affects the phenotype of the other homozygote (**ff**).

If two CF carriers had children what genotypes and phenotypes would we expect their children to have?

Predicting the genotypes of offspring

In the UK about 1 person in 25 is a cystic fibrosis carrier (**Ff**). They can pass the disease on to their children. When gametes are produced, each egg or sperm contains only one allele, in this case either **F** or **f**. These two types of gamete are produced in roughly equal numbers. The expected genotypes of children produced by two cystic fibrosis carriers can be shown in a genetic diagram, known as a Punnett square. This illustrates all the possible ways in which the two types of allele can combine, and thus shows the possible genotypes that can occur in the children.

Parents' phenotypes	normal	normal
Parents' genotypes	**Ff**	**Ff**
Gamete genotypes	**F** or **f**	**F** or **f**

		Gametes from mother	
		F	f
Gametes	F	FF	Ff
from father	f	Ff	ff

This means that every time a child is born to parents who are both carriers of cystic fibrosis, there is a 1 in 4 or 0.25 probability that the baby will have the genotype **ff** and suffer from cystic fibrosis. There is also a 1 in 4 probability that the genotype will be **FF**, and a 1 in 2 (2 in 4) or 0.5 probability that it will be **Ff**, a CF carrier.

Until very recently people with the genotype **ff** did not often survive to be adults and thus did not have any children. You might think that natural selection would have eliminated such a harmful allele. But in fact, heterozygotes (**Ff**) have some protection against the dangerous disease typhoid. So in areas where typhoid was common, carriers would have been at a definite advantage.

Q2.31 Give a genetic explanation for each of the three situations described at the top of page 90.

Q2.32 The genetic pedigree diagram on the right shows the inheritance of CF for three generations of one family. Explain how Frank could have inherited CF from James and Margaret, who do not have CF.

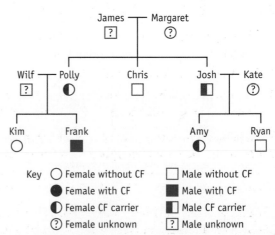

Key
○ Female without CF   □ Male without CF
● Female with CF   ■ Male with CF
◐ Female CF carrier   ◨ Male CF carrier
? Female unknown   ? Male unknown

Inheritance of some human genetic diseases

Cystic fibrosis is an example of **monohybrid inheritance**, so called because the characteristic is controlled by only one gene. Most human characteristics are inherited in a much more complex way, and are often influenced by environmental factors. However, there are a few characteristics controlled by single genes.

For example, **thalassaemia** is a genetic disease caused by recessive alleles of a gene on chromosome 11. The gene is involved in the manufacture of the protein haemoglobin, found in red blood cells, which carries oxygen around the body. A number of different mutations can affect this gene. Someone who is homozygous for one of these conditions either makes no haemoglobin at all, or makes haemoglobin that cannot carry out its function. The homozygous condition is often eventually lethal.

People who are heterozygous show no symptoms, but have some protection against malaria. For this reason thalassaemia is relatively common in people who live (or have ancestors who lived) in areas where malaria occurs or occurred in the past, particularly around the Mediterranean Sea. Because of this 'heterozygous advantage' the mutant alleles have not disappeared.

Other conditions such as **albinism**, **phenylketonuria** and **sickle cell anaemia** are also caused by single recessive alleles. **Achondroplasia**, on the other hand, is caused by a dominant allele. A homozygote, carrying two copies of the allele for achondroplasia, always dies. Someone who is heterozygous for this condition will show very restricted growth, usually attaining a height of about 125–130 cm. The trunk is of average length but the limbs are much shorter. Their mental powers are not affected. **Huntington's disease** and the ability to taste PTC (a bitter-tasting chemical called phenylthiocarbamide) are also caused by dominant alleles.

You should not suppose that characteristics determined by a single gene are found only in humans. They occur in all organisms. Indeed, they were first discovered in the garden pea by Gregor Mendel in the 1850s and 1860s.

The work of Mendel

Mendel is rightly known as the father of genetics. He was a monk and carried out a huge number of breeding experiments in the garden of his Moravian monastery. Mendel died without other scientists appreciating the significance of his work. However, he established that a number of characteristics of the garden pea were determined by separate genes. For example, whether the plants are tall (1.9–2.2 m in height) or short (0.3–0.5 m) is determined by one gene. Whether the seeds are smooth or wrinkled is determined by another gene.

Activity

Activity 2.17 lets you apply ideas about inheritance to some other situations. **A2.17S**

Activity

You can use Reebops to investigate inheritance in **Activity 2.16**. **A2.16S**

Checkpoint ✔

2.6 Produce a vocabulary list giving definitions of the following key genetic terms:

- gene
- locus
- homologous chromosomes
- alleles
- dominant allele
- recessive allele
- heterozygous
- homozygous
- carrier
- genotype
- phenotype.

2.7 How is cystic fibrosis treated?

The CF treatments currently available alleviate some of the symptoms. This can contribute to a longer life and improved quality of life, but as yet there is no known cure for cystic fibrosis.

? Did you know?

Treatments for CF

1 Medication

There are a wide range of medications commonly used to relieve the symptoms of CF. These include the following:

- Bronchodilators – these are drugs that are inhaled using a nebuliser. Air is blown through a solution to make a fine mist which is breathed into the lungs through a mouthpiece. The drugs relax the muscles in the airways, opening them up and relieving tightness of the chest.
- Antibiotics – the early diagnosis and treatment of lung infection is the cornerstone of CF treatment. Many antibiotics are used to kill or prevent the growth of bacteria in the lungs.
- DNAase enzymes – infection of the lungs leads to the accumulation of white blood cells in the mucus. The breakdown of these white blood cells releases DNA, which can add to the 'stickiness' of the mucus. DNAase enzymes can be inhaled using a nebuliser. They break down the DNA, so the mucus is easier to clear from the lungs.
- Steroids are used to reduce inflammation of the lungs.

2 Diet

Adults with CF are recommended to eat high-energy foods and their diet should include double the quantity of protein recommended for people who do not have CF. Some people with CF may also need salt supplements.

3 Digestive enzyme supplements

If the pancreatic duct is blocked, the food molecules in the small intestine cannot be broken down far enough to be absorbed. Taking enzyme supplements with food helps to complete the process of digestion.

4 Physiotherapy

Rhythmical tapping of the walls of the chest cavity (percussion therapy) can help loosen the mucus and improve the flow of air into and out of the lungs (Figure 2.48). Such treatment needs to be carried out regularly, twice a day.

5 Heart and lung transplant

If the lungs become badly damaged and very inefficient, other treatments may become ineffective at relieving the symptoms. The only option available may be to replace the damaged lungs with a heart and lung transplant.

Figure 2.48 Children learn how to perform their own physiotherapy. Here a child uses a flutter device. Exhaling through the flutter's special valve causes rapid changes in air pressure within the airways. The vibrations aid movement of the mucus.

Possible CF treatments for the future

Gene therapy

Current treatments only reduce some of the effects of the disease. Understanding the nature of the gene involved in cystic fibrosis has raised the possibility of a future cure through **gene therapy**. Effective gene therapy would treat the cause rather than the symptoms of the disease, but what is gene therapy?

In gene therapy the genotype and hence the phenotype of target cells (those affected by the disease) is altered. This is achieved as follows:

1 Normal alleles of the gene are inserted into the target cells, either using a genetically modified virus to infect the target cells or using liposomes (spherical phospholipid bilayers).

2 The normal form of the gene is transcribed and translated.

3 A functioning protein is produced in the target cells.

In the case of the CF gene, a functioning CFTR protein is produced and incorporated into the cell membrane, thus restoring the ion channel and avoiding the symptoms of CF.

How genes are inserted using viruses

In the virus, the DNA sequence that allows it to replicate is removed. This is replaced with the normal allele of the desired gene, along with a promoter sequence that initiates transcription and translation of the gene.

When we are infected with some viruses, the viral DNA becomes incorporated into our cells' own DNA. With other viruses the viral DNA remains independent within the nucleus of our cells. In trials with CFTR, the second type of virus is used.

The use of viruses is a potentially efficient form of gene transfer but it has been found to produce an inflammatory response, with treated patients experiencing symptoms such as headache, fatigue, fever and raised heart rate.

How genes are inserted using liposomes

First, a copy of the normal allele is inserted into a loop of DNA (called a **plasmid**). The plasmids are then combined with **liposomes** (spherical phospholipid bilayers). The positively charged head groups of the phospholipids combine with the DNA (a weak acid and so negatively charged) to form a liposome-DNA complex.

The CF patient breathes in an aerosol containing these complexes using a nebuliser. The liposomes fuse with epithelial cell membranes and carry the DNA into the cells (Figure 2.49 opposite).

Progress so far

Trials of CF gene therapy started in the UK back in 1993 and the 'normal' CFTR allele has been successfully transferred to the lung epithelial cells of CF patients. The presence of functioning ion channels is indicated by a reduction in the potential difference across the membranes in the nose and lungs, and increased chloride secretions. The results of trials have so far demonstrated a correction of the chloride secretion but not the associated sodium secretion problems.

In trial results published in 1999, chloride transport in the lungs was restored to 25% of normal. However, this type of improvement is temporary. Cells are continuously lost from the epithelium lining the airways so the transfer of the allele to these cells does not offer a permanent solution. The longest the correction has lasted in any trial

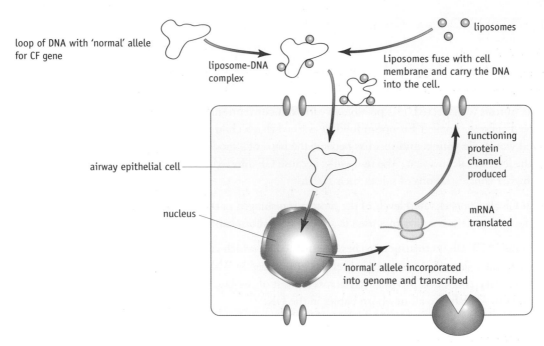

loop of DNA with 'normal' allele for CF gene

liposomes

liposome-DNA complex

Liposomes fuse with cell membrane and carry the DNA into the cell.

functioning protein channel produced

airway epithelial cell

mRNA translated

nucleus

'normal' allele incorporated into genome and transcribed

Figure 2.49 Gene therapy for cystic fibrosis may yet be successful, but difficulties with the delivery of the gene mean that the trials must continue.

is about 15 days, so this treatment would have to be repeated throughout the patient's life. Trials continue with the aim of improving the delivery of the gene and thus increasing the effectiveness of the treatment that may one day offer a cure.

Activity

In **Activity 2.18** complete the exercise on successes and failures in gene therapy and try the gene therapy crossword. **A2.18S**

The first condition in humans to have been successfully treated using gene therapy was a very rare disorder called severe combined immunodeficiency (SCID). Patients with this disease cannot make a particular enzyme needed for the immune system to work. In 1990, white blood cells were removed from a four-year-old girl with the disease, Ashanti DeSilva. The alleles for the functioning gene were inserted into her white cells using a virus. The cells were then replaced. Ashanti (now a young woman) has to have regular transfusions with the modified white cells but otherwise is fit and well. There were also reports of two Germans having been successfully treated for chronic granulomatous disease (CGD), a rare disease that prevents the body from killing bacteria and fungi, but one has since died.

It is hoped that gene therapy will help in the treatment of thalassaemia, haemophilia and sickle cell disease but to date only limited progress has occurred with these diseases. These treatments are all concerned with altering specific **somatic cells** (body cells) and are permitted under UK legislation. The alternative approach of altering the **germ cells** (sperm or eggs) so that every cell in the body contains the new gene is not permitted. There are ethical objections to such **germ line therapy** because of concerns about possible effects in future generations when the new gene is inherited.

2.8 Testing for CF

Even though 1 person in 25 in the UK is a CF carrier, the first time that most carriers realise that they carry the CF allele is when one of their children is born with the disease.

Once cystic fibrosis is suspected, it is possible to carry out conventional tests to confirm the diagnosis. Ancient European folklore warned that a child who tastes salty when kissed would die young, and this has become the basis of a modern test that measures the level of salt in sweat. The test works because CF-affected people have markedly higher concentrations of salt in their sweat.

People with CF also have elevated levels of the protein trypsinogen in their blood. A blood test for this protein is sometimes used in screening newborn babies for the disease.

Early diagnosis of CF allows treatment to begin immediately, which can improve health in later years. Testing of newborn babies was introduced in Wales in 1997, in Scotland in January 2003, and progressively across the rest of the UK. By April 2007 CF screening was available to all newborn babies in the UK.

Genetic testing

The gene that codes for the CFTR protein was sequenced in 1989. This led to the possibility of **genetic testing**, identifying the abnormal gene in the DNA of any cells. This in turn has paved the way for **genetic screening** to confirm the results of conventional tests, to identify carriers and also to diagnose CF in an embryo or foetus. (The terms 'genetic testing' and 'genetic screening' are sometimes distinguished but often, as here, used interchangeably.)

Genetic testing can be performed on any DNA, so it is possible to take samples of cheek cells, white blood cells or cells obtained from a foetus or embryo. The DNA is tested to see whether it contains the known base sequences for the most common mutations that cause cystic fibrosis.

Q2.33 The test is performed only for the most common mutations that cause CF. What does this mean about the reliability of the test? Will there be any false positives or negatives?

How can genetic screening be used?

To confirm a diagnosis

Genetic testing can confirm a diagnosis of CF. However, since there are a large number of different mutations of the CFTR gene which cause the disease, a negative result must be treated with caution. It is not currently feasible to test for all of the hundreds of possible mutations that lead to CF.

To identify carriers

Genetic testing can identify carriers. A sample of blood or cells taken from inside the mouth can be used to detect abnormal alleles in people without the disease who are heterozygous. Where there has been a history of cystic fibrosis in a family this can be of value in assessing the probability of having a child with the disease. Counselling is offered before and after testing, and parents can make informed decisions about how to proceed.

For testing embryos

Genetic testing can be used for prenatal testing of an embryo. Currently, there are two techniques used for obtaining a sample of cells from a child before birth. The more common is **amniocentesis**, which involves inserting a needle into the amniotic fluid to collect cells that have fallen off the placenta and foetus (Figure 2.50). This can be carried out at around 15–17 weeks of pregnancy, and involves a risk of causing a miscarriage of between 0.5% and 1%.

Chorionic villus sampling (CVS) is the second technique. Here, a small sample of placental tissue (which includes cells of the foetus) is removed, either through the wall of the abdomen or through the vagina. This can be carried out earlier, between 8 and 12 weeks, since there is no need to wait for amniotic fluid to develop. However, it carries a risk of about 1% to 2% of inducing a miscarriage.

Figure 2.50 A prenatal test during pregnancy such as amniocentesis or CVS involves techniques that carry some risk.

There are implications associated with the use of these tests which people should be aware of when deciding to have the tests so they can make informed choices. Both procedures present a risk of miscarriage to possibly healthy foetuses. If amniocentesis or CVS results in a positive genetic test for the disease, one possibility is for the woman to have an abortion. Having an abortion is easier for the woman, both physically and emotionally, in the earlier weeks of pregnancy. This can mean that the slightly higher risk of miscarriage associated with CVS is considered worthwhile.

Q2.34 Are there circumstances where people might choose not to test, even if the test was offered to them?

Testing before implantation (PIGD)

Genetic testing can enable pre-implantation genetic diagnosis (PIGD). When carrying out *in vitro* fertilisation (IVF), it is possible to test an embryo before it has implanted in the uterus. A cell can be removed from an embryo growing in culture when it has only 8 or 16 cells without harming the embryo (Figure 2.51). The DNA of the cell can be analysed and the results used to decide whether to place the embryo into the womb. IVF, however, is still an expensive and fairly unreliable procedure, and although this avoids the need for a possible abortion, this approach is not used routinely.

Figure 2.51 Why can one cell be removed at this stage without harming the developing embryo? We will see the answer to this question in Topic 3. Magnification ×110.

> **Activity**
>
> **Activity 2.19** looks at issues involved in using widespread carrier screening. **A2.19S**

Making ethical decisions – what is right and what is wrong?

How should we decide in life what is right and what is wrong? For example, should we always tell the truth? Can we ever justify turning down a request for help? Should Claire and Nathan have an abortion if they find that their unborn baby has cystic fibrosis?

All of us have **moral** views about these and other matters. For example, you might hold that lying and abortion are always wrong and helping people always right. But in order to maintain that something is ethically acceptable or unacceptable, you must be able to provide a reasonable explanation as to *why* that is the case.

There is no one universally accepted way of deciding whether something is ethically acceptable or not. What there are instead are a number of ethical frameworks, each of which allows you to work out whether a particular action would be right or wrong if you accept the ethical principles on which the framework is based. Usually you get the same answer whichever framework you adopt. But not always! This is why perfectly thoughtful, kind and intelligent people sometimes still disagree completely about whether a particular course of action is justified or not.

Ethical frameworks

We shall examine four widely used ethical frameworks. You should find these of value when considering various issues, such as genetic screening and abortion, raised in this topic. We will also refer to these frameworks in other topics in this course.

1 Rights and duties

Most of us tend to feel that there are certain human **rights** that should always be permitted. For example, we talk about the right to life, the right to a fair trial and the right to freedom of speech. Certain countries, for example the USA, have some of these rights enshrined in their constitutions.

If you have a right to something, then I may have particular **duties** towards you. For example, suppose that you are a six-month-old baby with a right to life and I am your parent. I have a duty to feed you, wash you, keep you warm and so on. If I don't fulfil these duties, I am failing to carry out my responsibilities and the police or social services may intervene.

But where do rights come from? Some people with a religious faith find them in the teachings of their religion. For example, the ten commandments in the Jewish scriptures talk about not stealing, not murdering, telling the truth and so on.

But nowadays, of course, many people, indeed in the UK most people, have little or no religious faith. So where can they – perhaps you – find rights? The simplest answer is that rights are social conventions built up over thousands of years. If you want to live in a society you have, more or less, to abide by its conventions.

2 Maximising the amount of good in the world

Perhaps the simplest ethical framework says that each of us should do whatever *maximises the amount of good in the world*. For example, should I tell the truth? Usually yes, as telling lies often ends up making people unhappy and unhappiness is not good. But sometimes telling the truth can lead to more unhappiness. If your friend asks you if you like the present they have just given you and you don't, would you tell the truth? Most of us would tell a 'white lie', not wanting to hurt their feelings.

This ethical approach is known as **utilitarianism**. Notice that utilitarians have no moral absolutes beyond maximising the amount of good in the world. A utilitarian would hesitate to state that anything is always right or always wrong. There might be circumstances in which something normally right (e.g. keeping a promise) would be wrong and there might be circumstances when something normally wrong (e.g. killing someone) would be right.

3 Making decisions for yourself

One of the key things about being a human is that we can make our own decisions. There was, for example, a time when doctors simply told their patients what was best for them. Now, though, there has been a strong move towards enabling patients to act autonomously. People act autonomously when they make up their own minds about things. If you have ever had an operation, you will probably have signed a consent form. The thinking behind this is that it isn't good for a surgeon to be allowed to operate on you unless you have given **informed consent**.

Of course, it is perfectly possible autonomously to decide to be absolutely selfish! A utilitarian would say we need to weigh the benefits of someone acting autonomously with any costs of them doing so. Only if the overall benefits are greater than the overall costs is **autonomy** desirable. Someone who believes in rights and duties might say that each of us has a right to act autonomously but also has a duty to take account of the effects of our actions on others.

4 Leading a virtuous life

A final approach is one of the oldest. This holds that the good life (in every sense of the term) consists of acting virtuously. This may sound rather old fashioned but consider the **virtues** that you might wish a good teacher/lecturer to have. She or he might be understanding, be able to get you to learn what you want or need to learn, and believe in treating students fairly.

Traditionally the seven virtues were said to be **justice**, prudence (i.e. wisdom), temperance (i.e. acting in moderation), fortitude (i.e. courage), faith, hope and charity. Precisely what leading a virtuous life means can vary and isn't always straightforward. Think about the virtues you might like to see in a parent, a doctor and a girl-/boyfriend. What would be the virtuous course of action for Claire and Nathan?

Genetic counselling

If a couple like Nathan and Claire are at risk of having a child with a genetic disorder, a genetic counsellor can provide advice. Such a person will help the couple understand how the disease is inherited and the chance that any child they conceive will have it. A genetic counsellor will explain the tests available and the possible courses of action, depending on the outcome of the tests. The counselling should help the couple decide such things as whether to be tested and whether to have children, use *in vitro* fertilisation with screening or use prenatal screening.

Q2.35 If both parents are found to be carriers, what are the options open to them?

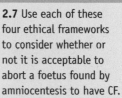

Checkpoint

2.7 Use each of these four ethical frameworks to consider whether or not it is acceptable to abort a foetus found by amniocentesis to have CF.

Activity

The role play in **Activity 2.20** lets you think about some of the issues covered in this topic. **A2.20S**

In **Activity 2.21** read a personal experience of another gene mutation to help you revisit some ideas considered earlier in the topic. **A2.21S**

Topic test

Now that you have finished Topic 2, complete the end-of-topic test before starting Topic 3.

Activity

Use **Activity 2.22** to check your notes using the topic summary provided. **A2.22S**

Voice of the genome

Why a topic called Voice of the genome?

Most multicellular organisms that can be seen around us, from apple trees to zebras, including magnolias, mice, monkeys, mushrooms, mayflies and us, *Homo sapiens*, start out in much the same way – as a very simple undifferentiated cell. But from that cell come scores of different cells, specialised for a huge variety of jobs and all produced in exactly the correct places. How is the amazing change from a single cell into a complex body achieved (Figure 3.1)? In this topic we follow the fate of the gametes produced by parents as these gametes start on the greatest journey of all – from separate cells to a full-grown adult.

Fertilised egg ⟶ early embryo of stem cells ⟶

specialised cells working together ⟶ complex multicellular organism

Figure 3.1 From a single cell to a full-grown adult.

William Harvey (1578–1657), one of the greatest anatomists who ever lived, said 'Everything comes from eggs'. But people used to think that the whole organism was wrapped up in miniature inside the sperm, ready to grow and be unfolded during development (Figure 3.2). Of course this isn't the case, but the truth is not so different. A fertilised egg contains a complete set of instructions. Inside every fertilised egg and body cell is a complete genome – DNA containing a full set of genes that control the growth and development of the whole organism.

Figure 3.2 It was once thought that a miniature baby was to be found inside the sperm cell.

In humans there are approximately 20 000 to 25 000 genes. We share 98.5% of our DNA sequence with chimpanzees, making them our closest relatives in the animal kingdom. However, we also share 50% of our genes with bananas, which is food for thought! Genes provide the instruction manual to build and control not only a single cell but also the great range of specialised cells that make up an adult organism. How do these genes talk to the different cells?

In some organisms, features you might think are controlled exclusively by genes are not quite so straightforward, with genes and the environment interacting. What is the relationship between genes and the environment? Is cancer inherited, caused by our environment, or a combination of both?

Overview of the biological principles covered in this topic

Gamete structure and function provide the starting point for this topic. You will look at their role in fertilisation. To understand how a single cell divides and grows into a whole organism, you need a good grasp of cell structure and a detailed knowledge of the cell cycle. At GCSE the basics of cell division were covered; here you will gain a more detailed picture of this process, including the role of both mitosis and cytoplasmic division.

You will discover how the first cells produced have the potential to develop into any part of the body, and see how these stem cells are becoming increasingly important to medical research. They offer a huge potential for the creation of new treatments and therapies. But these new developments are controversial, and you will have to consider the arguments for and against the use of these stem cells in medicine to clarify your own views on the issue.

You will learn how cells become specialised through differential gene expression and how individual specialised cells work together as tissues, organs and organ systems.

You will also look at how the characteristics of an organism are controlled by both genetic make-up and the environment.

Review

Are you ready to tackle Topic 3 *Voice of the genome*?

Complete the GCSE review and GCSE review test before you start.

3.1 In the beginning

Have you ever tried incubating an egg bought from the supermarket (Figure 3.3)? If so it is unlikely that the egg hatched because these eggs are unfertilised. The hens they come from are kept in female-only flocks; no cockerels are present so no sperm have combined with these eggs. To get a chick you need a fertilised egg that divides and differentiates into a complex multicellular body.

The life cycle of most multicellular organisms starts in this way. Gametes combine to form a single fertilised egg that develops into a new adult made up of numerous specialised cells organised into tissues and organs. But are all the cells basically the same as that fertilised egg?

Figure 3.3 Which came first?

Key biological principle: Are all cells basically the same?

Cells do have many common features, but it is possible to distinguish two basic types: **prokaryotic** and **eukaryotic** cells.

Prokaryotic cells

Bacteria and cyanobacteria (photosynthetic bacteria) together make up the Prokaryotae Kingdom (see Topic 4). Their cells do not have nuclei or other membrane-bound cell organelles (Figures 3.4 and 3.5). This type of cell is called a 'prokaryotic cell', meaning 'before the nucleus'.

Figure 3.4 Electron microscope image of the bacterium *Vibrio cholerae*. Magnification ×36 000.

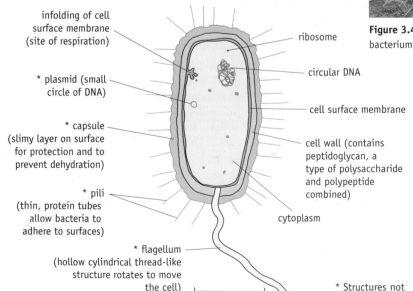

infolding of cell surface membrane (site of respiration)

ribosome

circular DNA

* plasmid (small circle of DNA)

cell surface membrane

* capsule (slimy layer on surface for protection and to prevent dehydration)

cell wall (contains peptidoglycan, a type of polysaccharide and polypeptide combined)

* pili (thin, protein tubes allow bacteria to adhere to surfaces)

cytoplasm

* flagellum (hollow cylindrical thread-like structure rotates to move the cell)

0.5–5 µm

* Structures not always present

Figure 3.5 The structure of a generalised bacterial cell.

CONTINUED ▶

Most prokaryotes are extremely small with diameters between 0.5 and 5 µm. Their DNA is not associated with any proteins and lies free in the cytoplasm. A cell wall is always present in prokaryotic cells.

Eukaryotic cells

All other living organisms have cells that contain discrete membrane-bound organelles such as nuclei, mitochondria and chloroplasts (plants only). These are 'eukaryotic cells', meaning 'true nucleus'. Eukaryotic cells are larger than prokaryotic cells with diameters of 20 µm or more. Unlike the prokaryotes, not all eukaryotic cells have a cell wall. Organisms with eukaryotic cells are eukaryotes.

Inside the eukaryotic animal cell

The classical child's diagram of a cell resembles a fried egg: one circle with a large blob in the middle. Using a simple light microscope this is almost as much detail as it is possible to see. However, the use of electron microscopy reveals a wealth of structures within the cell, often described as the cell ultrastructure (Figure 3.6).

> **Activity**
>
> In **Activity 3.1** you look in more detail at the relationship between the three-dimensional structures and the functions of organelles in the cell using the interactive cell. **A3.01S**

Figure 3.6 Electron micrograph of a pancreas cell shows the cell's ultrastructure. It contains many mature secretory vesicles (dark round structures) and immature secretory vesicles (paler round structures).

Figure 3.7 shows a two-dimensional representation of the cross-section of a generalised animal cell. You can see there are many structures within the cell cytoplasm. These are shown in three dimensions in Figure 3.8. Use these diagrams to answer the questions below.

Q3.1 Follow a straight line across the cell in Figure 3.7 from A to B and identify each of the structures that the line crosses using Figure 3.8 (overleaf).

Q3.2 Identify two structures within the cell which are not crossed by this line.

The structures and functions of plant cells will be studied in Topic 4.

Figure 3.7 A generalised animal cell.

CONTINUED ▶

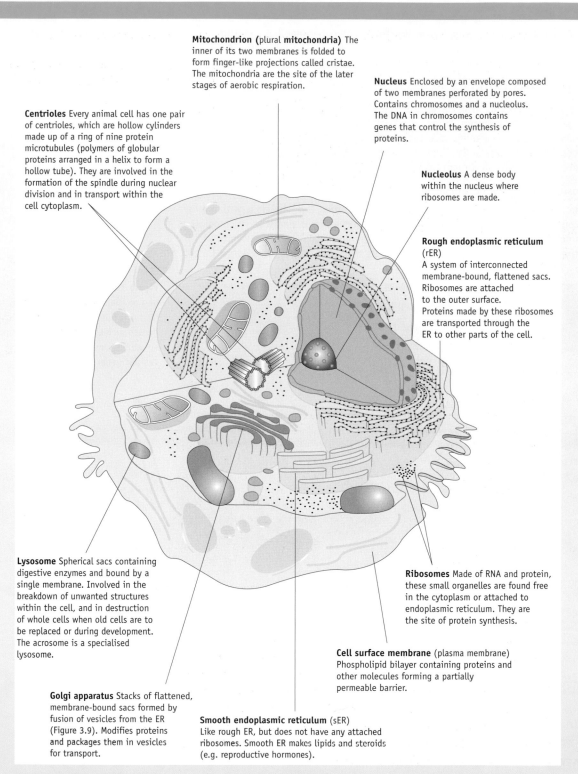

Mitochondrion (plural **mitochondria)** The inner of its two membranes is folded to form finger-like projections called cristae. The mitochondria are the site of the later stages of aerobic respiration.

Nucleus Enclosed by an envelope composed of two membranes perforated by pores. Contains chromosomes and a nucleolus. The DNA in chromosomes contains genes that control the synthesis of proteins.

Centrioles Every animal cell has one pair of centrioles, which are hollow cylinders made up of a ring of nine protein microtubules (polymers of globular proteins arranged in a helix to form a hollow tube). They are involved in the formation of the spindle during nuclear division and in transport within the cell cytoplasm.

Nucleolus A dense body within the nucleus where ribosomes are made.

Rough endoplasmic reticulum (rER) A system of interconnected membrane-bound, flattened sacs. Ribosomes are attached to the outer surface. Proteins made by these ribosomes are transported through the ER to other parts of the cell.

Lysosome Spherical sacs containing digestive enzymes and bound by a single membrane. Involved in the breakdown of unwanted structures within the cell, and in destruction of whole cells when old cells are to be replaced or during development. The acrosome is a specialised lysosome.

Ribosomes Made of RNA and protein, these small organelles are found free in the cytoplasm or attached to endoplasmic reticulum. They are the site of protein synthesis.

Cell surface membrane (plasma membrane) Phospholipid bilayer containing proteins and other molecules forming a partially permeable barrier.

Golgi apparatus Stacks of flattened, membrane-bound sacs formed by fusion of vesicles from the ER (Figure 3.9). Modifies proteins and packages them in vesicles for transport.

Smooth endoplasmic reticulum (sER) Like rough ER, but does not have any attached ribosomes. Smooth ER makes lipids and steroids (e.g. reproductive hormones).

Figure 3.8 A three-dimensional representation of a generalised animal cell.

CONTINUED ▶

Q3.3   Look at the electron micrograph in Figure 3.9 below and identify the structures labelled **A** to **E**.

Figure 3.9 Electron micrograph of part of a cell from the pancreas of the bat *Myotis lucifugus*. The cell produces pancreatic juice and enzymes. Magnification ×1300.

Checkpoint

3.1 Draw up a table to compare prokaryotes and eukaryotes.

 Did you know?

The origins of chloroplasts and mitochondria

Chloroplasts and mitochondria are thought to have originally been independent prokaryotes. Very early in the evolution of multicellular organisms, they developed a mutualistic relationship (of benefit to both partners) with eukaryotic cells. Perhaps these ancestors of chloroplasts and mitochondria gained protection and a more constant environment by being inside eukaryotes. The eukaryotes probably benefited from the products of the metabolism of the prokaryotes: sugar and oxygen from the photosynthetic prokaryotes that became chloroplasts, and ATP from the non-photosynthetic prokaryotes that became mitochondria. To this day, both chloroplasts and mitochondria contain their own DNA separate from the DNA of the cells they inhabit. However, they can no longer replicate themselves independently but need to rely partly on enzymes made by their hosts.

Cells are dynamic

The cell is not static; there is continual movement of molecules within it. In Topic 2 you saw the movement of RNA from the nucleus to the ribosomes for protein synthesis. Once the proteins have been synthesised they are processed and move through the cell to where they are needed. In addition, many proteins such as enzymes, hormones and signal proteins are released from the cells where they are made. This movement of proteins through the cell involves the endoplasmic reticulum, Golgi apparatus and vesicles. These membrane structures are continually created and lost within the cell, as outlined in Figure 3.10 overleaf.

 Activity

You can see the protein movement within the cell in the animation that accompanies **Activity 3.2. A3.02S**

nucleus

1 Transcription of DNA to mRNA.

2 mRNA leaves nucleus.

nuclear envelope

3 Protein made on ribosomes enter rough ER.

4 Protein moves through the ER assuming three-dimensional shape en route.

5 Vesicles pinched off the rough ER contain the protein.

6 Vesicles from rough ER fuse to form the flattened sacs of the Golgi apparatus.

8 Vesicles pinched off the Golgi apparatus contain the modified protein.

7 Proteins are modified within the Golgi apparatus.

secretory vesicle

cell surface membrane

9 Vesicle fuses with cell surface membrane releasing protein, such as extracellular enzymes.

Figure 3.10 The production of proteins and their route through the cell.

Getting started

All eukaryotic cells share the same basic structures and many functions. As multicellular organisms develop, cells take on specialised roles. Gametes, the starting point in the story of development, are highly specialised, with structures and functions that are obviously different from other body cells.

Gametes

Mammalian gametes

You will remember from GCSE that in mammals the **sperm** and **ovum** (egg) cells are the **gametes** or sex cells (Figure 3.11).

The gametes are adapted for their roles in sexual reproduction. In humans, as in other mammals, the ovum is a large cell incapable of independent movement. It is wafted along one of the oviducts from the ovary to the uterus (Figure 2.31A, page 80) by ciliated cells lining the tubes and by muscular contractions of the tubes. The cytoplasm of the ovum contains protein and lipid food reserves for a developing embryo. Surrounding the cell is a jelly-like coating called the zona pellucida.

The sperm cell is much smaller than the ovum and is motile (can move). To enable it to swim, the sperm cell has a long tail powered by energy released by mitochondria. Males continually produce large numbers of sperm once they have reached puberty. Sperm which enter the vagina during intercourse swim through the uterus, their passage assisted by muscular contractions of the uterus walls. If intercourse takes place at about the time of ovulation sperm may meet the ovum in the oviduct. The sperm are attracted to the ovum by chemicals released from it. To penetrate the ovum, the **acrosome** in the head of the sperm releases digestive enzymes, which break down the zona pellucida of the ovum (Figure 3.12). The acrosome is a type of lysosome. Lysosomes are enzyme-filled sacs found in the cytoplasm of many cells. They are involved in the breakdown of unwanted structures within the cell.

Checkpoint

3.2 Look at Figure 3.11. Make a list of the differences between the mammalian gamete cells, and explain the value of each feature listed.

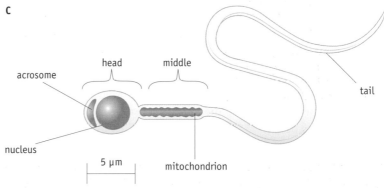

Once a sperm fuses with and penetrates the membrane surrounding the egg, chemicals released by the ovum cause the zona pellucida, to thicken, preventing any further sperm entering the egg. The sperm nucleus that enters the egg fuses with the egg nucleus to produce a fertilised egg.

Figure 3.11 A Scanning electron micrograph of sperm on an ovum during fertilisation. Magnification ×1100. Diagram of **B** a human ovum and **C** a human sperm.

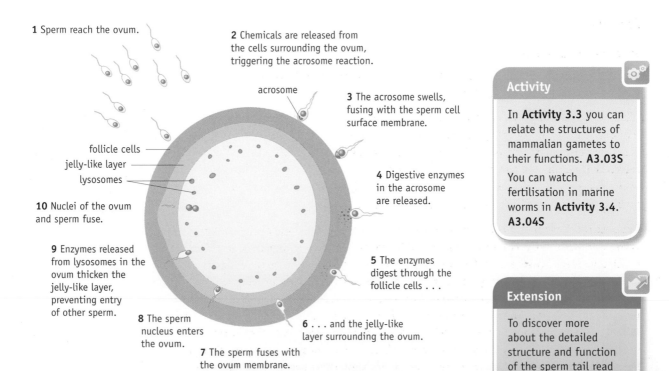

Activity

In **Activity 3.3** you can relate the structures of mammalian gametes to their functions. **A3.03S**

You can watch fertilisation in marine worms in **Activity 3.4**. **A3.04S**

Extension

To discover more about the detailed structure and function of the sperm tail read **Extension 3.1. X3.01S**

Figure 3.12 When sperm meets ovum, enzymes released from the acrosome digest through the outer coating of the ovum.

Plant gametes

The male part of the plant is the stamen, made up of the anther attached to a stalk known as a filament. Inside the anther, cells divide to produce pollen grains, which contain the male gamete nuclei. Inside the female ovary, one or more ovules develop which contain the female gametes, the ova (singular ovum) (Figure 3.17).

Gamete cells are unusual

Although the number of chromosomes differs from species to species, all 'normal' individuals within a species will have the same number of chromosomes in each cell. Human cells contain 46 chromosomes made up of 22 homologous pairs and one pair of sex chromosomes; fruit fly cells contain 8 chromosomes made up of three homologous pairs and one pair of sex chromosomes; white campion plant cells have 24 chromosomes made up of 11 homologous pairs and one pair of sex chromosomes; rice cells contain 24 chromosomes made up of 12 homologous pairs (no sex chromosomes as separate male and female individuals are not produced); and so on.

The fundamental difference between gametes and other cells is the number of chromosomes they contain. In Topic 2 you saw that gametes have half the number of chromosomes found in normal cells – one chromosome from each homologous pair – but have you considered why this is? Think about what happens at fertilisation. If the sperm and ovum cells both had the full chromosome number, which is 46 in humans, then the zygote would have 92 chromosomes. When this individual reproduced the zygote would contain 184 chromosomes. With each generation the number of chromosomes would double, which obviously wouldn't work. This is why the gametes contain half the full number of chromosomes, 23 in humans, made up of one of each homologous pair and one sex chromosome. When the gametes fuse, the full number of 46 is restored as you can see in Figure 3.13.

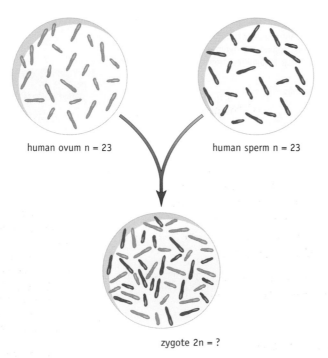

human ovum n = 23   human sperm n = 23

zygote 2n = ?

Figure 3.13 Haploid gametes come together to form a diploid zygote. What number should replace the question mark? Count them if you are not sure! The number of chromosomes in gametes and zygotes varies from species to species. Human zygotes contain 46 chromosomes, dogs have 78 and peas have 14.

How do gametes form?

There are two different types of cell division in living organisms. One involves **mitosis** and produces new body cells as an organism grows and develops. This retains the full number of chromosomes, called the **diploid** number (2*n*) (46 in humans). The other type of cell division involves **meiosis** and produces gametes with only half the number of chromosomes, called the **haploid** number (*n*) (23 in humans). This form of cell division occurs in the ovaries and testes of animals, and the ovaries and anthers of flowering plants. This type of cell division is shown in Figure 3.14.

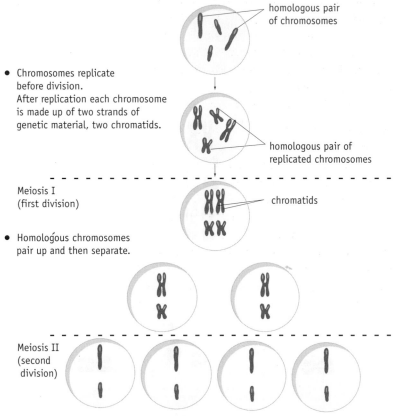

- Chromosomes replicate before division.
  After replication each chromosome is made up of two strands of genetic material, two chromatids.

Meiosis I (first division)

- Homologous chromosomes pair up and then separate.

Meiosis II (second division)

- Chromatids separate and gametes are formed, each with half the original number of chromosomes.

Figure 3.14 Gamete production by meiosis.

In female plants only one of the four nuclei at the end of meiosis forms an ovum – the others disintegrate. In male plants the four haploid cells produced at the end of meiosis undergo another division, this time by mitosis, where each nucleus divides to give two haploid nuclei within the pollen grain. One is the tube nucleus which controls the development of the pollen tube, and the second is known as the generative nucleus which will divide again to form two gamete nuclei (Figure 3.17).

Meiosis has two important roles in biology. Firstly it results in haploid cells, which are necessary to maintain the diploid number after fertilisation. Secondly, it helps create genetic variation among offspring.

How does meiosis result in genetic variation?

The shuffling of existing genetic material into new combinations during meiosis is important in creating genetic variation. This shuffling includes: **independent assortment** and **crossing over**.

Extension

To find out how the pollen grain and embryo sac are actually separate haploid generations within the life cycle of the plant, read **Extension 3.2** on alternation of generations. **X3.02S**

Independent assortment

During meiosis only one chromosome from each pair ends up in each gamete. The independent assortment of the chromosome pairs as they line up during meiosis I is a source of genetic variation. This process is random; either chromosome from each pair could be in any gamete. This way of sharing out chromosomes produces genetically variable gametes. The way that this happens is shown in Figure 3.15. An organism with six chromosomes, that is three homologous pairs X**X**, Y**Y** and Z**Z**, could form eight combinations in its gametes. When these join with another set at fertilisation, this pretty well guarantees that individuals produced from sexual reproduction are genetically different from each other. In organisms with many chromosomes, such as humans with 23 pairs, the number of possible combinations of the chromosomes is so large that it is unlikely any two siblings will have the same genetic make-up unless they are identical twins.

The arrangement of each chromosome pair during the first division in meiosis is completely random. In a cell with three chromosome pairs both the arrangements shown here are possible, for example.

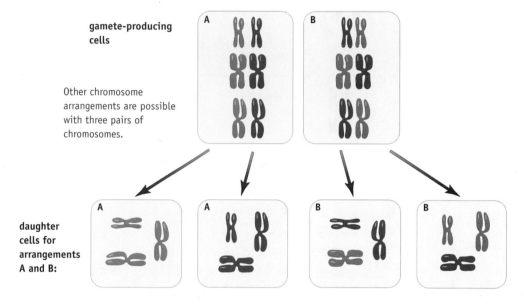

The daughter cells of the first meiotic division contain different assortments of chromosomes depending on the alignment of the chromosomes during the first meiotic division.

Figure 3.15 Independent assortment introduces variation.

Q3.4 Give the possible combinations of chromosomes that could occur in the gametes of an organism with three homologous pairs X**X**, Y**Y** and Z**Z**.

Q3.5 How many possible combinations of maternal and paternal chromosomes could be found in the gametes of organisms with $2n = 8$, and organisms with $2n = 10$?

Crossing over

During the first meiotic division, homologous chromosomes come together as pairs and all four **chromatids** come into contact. At these contact points the chromatids break and rejoin, exchanging sections of DNA (see Figure 3.16). The point where the chromatids break is called a **chiasma** (plural **chiasmata**), and several of these often occur along the length of each pair of chromosomes, giving rise to a large amount of variation.

> **Checkpoint** ✔
>
> **3.3** Produce a concept map or table which summarises how genetic variation is generated.

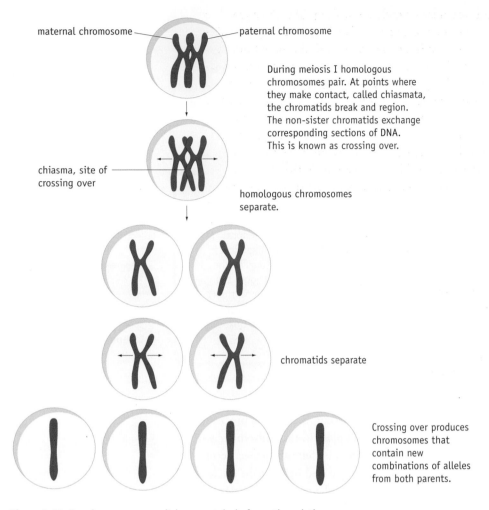

maternal chromosome

paternal chromosome

During meiosis I homologous chromosomes pair. At points where they make contact, called chiasmata, the chromatids break and region. The non-sister chromatids exchange corresponding sections of DNA. This is known as crossing over.

chiasma, site of crossing over

homologous chromosomes separate.

chromatids separate

Crossing over produces chromosomes that contain new combinations of alleles from both parents.

Figure 3.16 Crossing over can result in a great deal of genetic variation.

Q3.6 What would the chromosomes in the gametes produced from the homologous chromosomes in cells A and B below look like after crossing over has occurred?

A   B

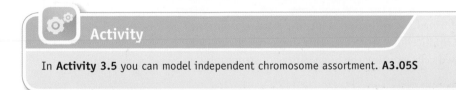

Activity

In **Activity 3.5** you can model independent chromosome assortment. **A3.05S**

Getting together

Fertilisation in mammals

To produce a new individual, the nuclei from the gametes have to combine in the process of **fertilisation** (Figure 3.12). In mammals the nucleus from one sperm enters the ovum, and the genetic material of the ovum and sperm fuse, forming a fertilised ovum called a **zygote**. This cell now contains genetic material from both parents.

? Did you know?

One cause of male infertility

For the human zygote to develop, the gamete nuclei have to fuse and a chemical from the sperm cytoplasm is required to activate the fertilised cell. This chemical is a protein called oscillin. It causes calcium ions to move in and out of stores in the cytoplasm of the ovum. These oscillations of calcium ion concentration trigger the zygote to begin developing into an embryo. Oscillin is concentrated in the first part of the sperm to attach to the ovum, and enters before the male nucleus in order to activate the ovum. It is thought that low levels of oscillin in sperm may be linked to male infertility, and this is a current area of research.

Extension

To find out about sperm competition read **Extension 3.3. X3.03S**

Fertilisation in flowering plants

In flowering plants, nuclei from the gametes also have to combine in the process of fertilisation (Figures 3.17 and 3.18). Fertilisation takes place in the embryo sac within the ovule. The pollen grain germinates on the style and a pollen tube grows down through the style towards the ovary, with its growth controlled by the tube nucleus. The pollen grain contains two nuclei, the tube nucleus and a generative nucleus. On germination of the pollen, the generative nucleus divides to form two haploid gamete nuclei which move down the pollen tube. The tube grows through a microscopic pore into the embryo sac and the two male gamete nuclei enter the sac. One fuses with the egg cell and forms a diploid zygote and the second fuses with two nuclei in the embryo sac called polar nuclei to form a triploid cell.

The diploid zygote divides to form the **embryo**. The triploid cell divides to form the seed's storage tissue, endosperm.

Activity

In **Activity 3.6** you view the growth of pollen tubes down the style. **A3.06S**

For most organisms the diploid cell formed at fertilisation is just the starting point on the road to creating a more complex multicellular structure. This single cell divides and gives rise to numerous more specialised cells, creating the huge variety of structures within a body.

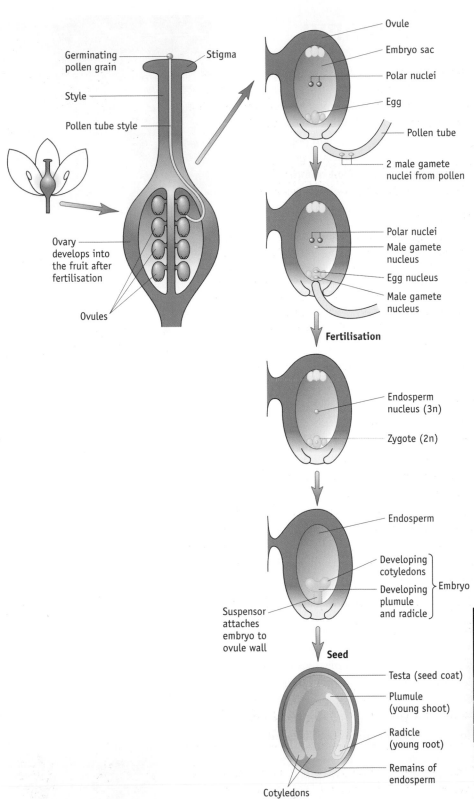

Germinating pollen grain

Stigma

Style

Pollen tube style

Ovary develops into the fruit after fertilisation

Ovules

Ovule

Embryo sac

Polar nuclei

Egg

Pollen tube

2 male gamete nuclei from pollen

Polar nuclei

Male gamete nucleus

Egg nucleus

Male gamete nucleus

Fertilisation

Endosperm nucleus (3n)

Zygote (2n)

Endosperm

Developing cotyledons

Developing plumule and radicle

Embryo

Suspensor attaches embryo to ovule wall

Seed

Testa (seed coat)

Plumule (young shoot)

Radicle (young root)

Remains of endosperm

Cotyledons

Figure 3.17 Double fertilisation involving two male gametes is unique to flowering plants.

Figure 3.18 Scanning electron micrograph of germinating pollen grains. Pollen tubes are visible growing into the stigma. Magnification ×220.

3.2 From one to many: the cell cycle

Elephants (Figure 3.19) have much bigger bodies than mice. However, elephant cells are little or no larger than the corresponding mouse cells – there are just far more of them. Elephants, mice, trees and poppies all start as a single fertilised egg with the potential to divide and grow into the complete body of a new individual. To make the vast number of cells required to build individuals, new cell contents must be synthesised and then one cell must divide into two. This well-organised pattern of events is called the **cell cycle**.

Figure 3.19 How many cells are there in an elephant? Humans have approximately ten thousand thousand mi (10^{13}) cells.

The cell cycle

The cell cycle can be divided into two distinct parts: interphase and division (Figure 3.20).

Preparation for division: interphase

Interphase is a time of intense and organised activity during which the cell synthesises new cell components such as organelles and membranes, and new DNA. The formation of new cellular proteins occurs throughout interphase, whereas DNA synthesis occurs during the **S** (for synthesis of DNA) phase. The S phase separates the first gap or **G1** phase from the second gap or **G2** phase, as shown in Figure 3.20.

The length of interphase differs depending on the role of the cell. In the developing human embryo, there is no interphase for the first few divisions – the zygote already contains the materials needed to form the first 16 or so cells. In these first few divisions, the embryo divides without growing in size, producing smaller cells with each cell cycle. This makes the embryonic cell cycle much faster than those of other body cells.

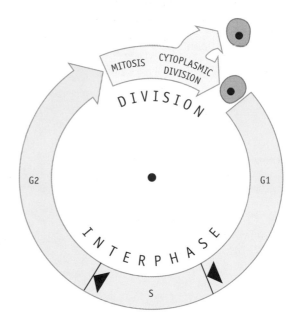

Figure 3.20 The stages in the cell cycle.

The S and G2 phases of most cells remain relatively constant in duration. The length of the G1 phase is more variable; some cells can take weeks, months or even years to complete this phase. For example, liver cells may divide only once every one or two years due to an extended G1 phase. In seeds, the cells of the embryo usually remain in the G1 phase until germination occurs. Some cells such as nerve and muscle cells never divide again, remaining permanently in a non-dividing state. In plants, cell division is localised in small groups of cells called meristems, with most other cells not dividing.

Q3.7 Suppose the cell cycle shown in Figure 3.20 lasts approximately 24 hours. Calculate the approximate length of time that the G1, S, G2 and division phases will last in this cell cycle.

Q3.8 Look at Figure 3.21. What do you notice about the nucleus during interphase? What does this suggest to you about what may be going on inside the cell?

Figure 3.21 The lower two cells are typical of interphase.

The interphase nucleus is a fairly uniform, featureless structure with one or two darker-staining regions called nucleoli (singular nucleolus). Ribosomes are formed in the nucleoli, and they give nucleoli their dark appearance in electron micrographs. Ribosomes are made of protein and rRNA (ribosomal RNA). The rest of the nucleus contains the chromosomes, which are made up of DNA associated with proteins (Figure 3.22).

During interphase, the individual chromosomes are unravelled (Figure 3.22). This allows access to the genetic material enabling new proteins to be synthesised. In preparation for cell division, the cell synthesises additional cytoplasmic proteins and organelles. The cell must also produce copies of DNA for the two new cells. It is vital that this DNA is identical in both structure and quantity to the DNA in the original cell. This is achieved by DNA replication, as described in Topic 2, page 88.

Q3.9 Which organelles would you expect to be most active in a cell that is producing large quantities of new proteins?

DNA double helix.

2 nm

DNA winds around proteins.

11 nm

proteins

DNA and proteins coil to form chromatin fibre.

30 nm

protein scaffold

Chromatin fibre attaches to a protein scaffold forming loops.

300 nm

700 nm

Folding the protein scaffolding produces the condensed chromosome structure seen during nuclear division.

1400 nm

Figure 3.22 Organisation of DNA in chromosomes.

Cell division

By the end of interphase, the cell contains enough cytoplasm, organelles and DNA to form two new cells. The next step is to share out both the DNA and the contents of the cytoplasm so that each new cell can function independently. The DNA is separated in **nuclear division** (mitosis). **Cytoplasmic division** follows this.

Cell division is a continuous process, where a single cell with double the usual amount of cell contents becomes two new cells. However, it is possible to describe four stages during nuclear division by the behaviour of the chromosomes and other structures within the cell. Figure 3.23 shows what is happening in each of these stages of mitosis, known as **prophase**, **metaphase**, **anaphase** and **telophase**.

Activity

In **Activity 3.7** try to work out what happens in each of the four stages of mitosis using the mitosis flick book. **A3.07S**

Compare this with the cell cycle/mitosis animation in **Activity 3.8**. **A3.08S**

During **interphase**, new cell organelles are synthesised and DNA replication occurs. By the end of interphase, the cell contains enough cell contents to produce two new cells.

Interphase.

After nuclear division, the final reorganisation into two new cells occurs. This is called **cytoplasmic division**. In animal cells, the cell surface membrane constricts around the centre of the cell. A ring of protein filaments bound to the inside surface of the cell surface membrane is thought to contract until the cell is divided into two new cells. It has been proposed that the proteins actin and myosin, responsible for muscle contraction, may also be the proteins responsible for cytoplasmic division. Instead of undergoing this constriction, plant cells synthesise a new cell plate between the two new cells (Figure 3.24 overleaf).

Cytoplasmic division.

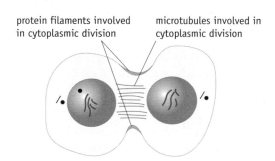

protein filaments involved in cytoplasmic division   microtubules involved in cytoplasmic division

4 Telophase.

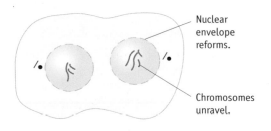

Nuclear envelope reforms.

Chromosomes unravel.

This last stage of mitotic division is called **telophase**. This is effectively the reverse of prophase. The chromosomes unravel and the nuclear envelope reforms, so that the two sets of genetic information become enclosed in separate nuclei.

Figure 3.23 Follow the sequence of events during cell division, clockwise in the direction of the arrows. The diagrams show a generalised animal cell containing only four chromosomes.

1 Prophase.

Chromosomes condense.

Spindle forms.

Nuclear envelope breaks down forming vesicles in the cytoplasm.

During **prophase**, the chromosomes condense, becoming shorter and thicker, with each chromosome visible as two strands called **chromatids**. Apart from the occasional mutation, the two strands are identical copies of one another, produced by replication. They are effectively two chromosomes joined at one region called the **centromere** (Figure 3.25 overleaf).

Q3.10 Why do the chromosomes condense?

During prophase, microtubules from the cytoplasm form a three-dimensional structure called the **spindle**. The **centrioles** move around the nuclear envelope and position themselves at opposite sides of the cell. These form the two poles of the spindle, and are involved in the organisation of the spindle fibres. The spindle fibres form between the poles. The widest part of the spindle is called the equator.

The breakdown of the nuclear envelope signals the end of prophase and the start of **metaphase**. The chromosomes' centromeres attach to spindle fibres at the equator. When this has been completed the cell has reached the end of metaphase.

3 Late anaphase.

In the next stage of mitosis, **anaphase**, the centromeres split. The spindle fibres shorten, pulling the two halves of each centromere in opposite directions. One chromatid of each chromosome is pulled to each of the poles. Anaphase ends when the separated chromatids reach the poles and the spindle breaks down.

2 Metaphase.

Q3.11 Why does the nuclear envelope disintegrate?

Spindle fibres shorten, pulling chromatids towards the poles.

Chromosomes move to equator.

Centromeres attach to spindles.

Activity

View the different stages of mitosis in root tip squashes and determine the length of each stage in **Activity 3.9. A3.09S**

You can also count cells at different stages or use the web tutorial in **Activity 3.10. A3.10S**

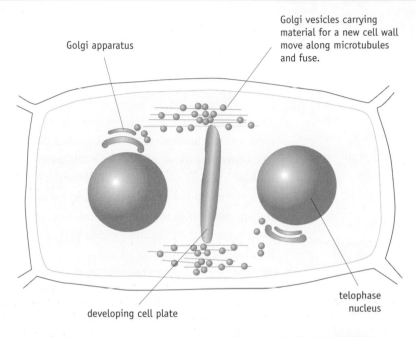

Golgi apparatus

Golgi vesicles carrying material for a new cell wall move along microtubules and fuse.

developing cell plate

telophase nucleus

Figure 3.24 A plant cell undergoing cytoplasmic division. Photo magnification ×4000.

homologous chromosomes

centromere

chromatids

Figure 3.25 During prophase, each chromosome becomes visible as two strands, often called sister chromatids. The centromere joining the two chromatids may be in the middle of the strands or towards one end.

✔ Checkpoint

3.4 a Draw a sequence of annotated cells which clearly show what happens during each stage of mitosis.

b Highlight the key features of mitosis that ensure identical daughter cells are produced.

Q3.12 Look at Figure 3.26 and compare the number of cells in telophase with the number in anaphase. What does this suggest to you about the relative lengths of time that these two phases take?

Figure 3.26 Cells undergoing mitosis. Magnification ×1300.

Key biological principle: Why is mitosis so important?

Mitosis ensures genetic consistency, with daughter cells genetically identical to each other and to the parent cell, each containing exactly the same number and type of chromosomes as their parent cell. This is achieved by:

- DNA replication prior to nuclear division
- the arrangement of the chromosomes on the spindle and the separation of chromatids to the poles.

This genetic consistency is important in growth and repair, and also in asexual reproduction.

Growth and repair

Mitosis occurs in the growth of any organism as it develops from a single cell into a multicellular organism. It ensures that a multicellular organism has genetic consistency; all the cells in the body have the same genetic information. Some organisms can regenerate lost or damaged parts of their bodies using mitosis, and starfish can even regrow a completely new body from a fragment (Figure 3.27A). Mitosis also allows old and damaged cells to be replaced with identical new copies.

Asexual reproduction

Many organisms reproduce without producing gametes. They grow copies of themselves by mitosis, producing offspring that are genetically identical to each other and to their parent. This form of reproduction is asexual, meaning 'without sex'.

Asexual reproduction occurs when bacteria undergo binary fission; the bacterial cell grows and then divides into two new cells. *Hydra* reproduce asexually by budding (Figure 3.27B); an outgrowth from the parent detaches to form a new individual. The single-celled fungus yeast also reproduces in this way. Asexual reproduction is common in plants too, where it is sometimes known as vegetative reproduction. The growth of new plants from tubers and bulbs are familiar examples (Figure 3.27C).

It is worth noting that there are organisms, such as mosses, liverworts and all plants, that reproduce asexually at one stage in their life cycle and sexually at another.

Figure 3.27 A Many organisms can regenerate parts of their bodies using mitosis. The crown-of-thorns starfish eats coral and has caused considerable damage to some coral reefs including the Great Barrier Reef off Australia. Early attempts to control this starfish included chopping them up and throwing the bits back into the sea. Unfortunately, due to their ability to regenerate new individuals from many of the pieces, this just increased their numbers. **B** Freshwater *Hydra* attached to water weed. Notice the young hydra that is budding from the parent by asexual reproduction. Eventually constriction at the point of attachment causes the release of the young hydra as a separate individual. **C** A hyacinth bulb is a highly modified underground shoot with fleshy leaves. Each of the shoot's buds can develop into a new plant.

CONTINUED ▶

Q3.13 For each pair of statements, decide which one refers to asexual and which to sexual reproduction:

a one parent needed/two parents needed

b gametes/no gametes

c fertilisation/no fertilisation

d offspring show genetic variation/offspring genetically identical

e allows for rapid increase in population numbers/ allows only slow increase in numbers

f unfavourable changes in the environment usually kill whole population/changes in the environment may only kill poorly adapted individuals

g effective for rapid colonisation of new areas/enables adaptation to changes in the environment.

Extension

To find out more about asexual reproduction that involves meiosis, read **Extension 3.2** on alternation of generations. **X3.02S**

Early embryonic development – stem cells

Cells in the early embryo

After a human zygote has undergone three complete cell cycles, it consists of eight identical cells. Each of these cells is said to be **totipotent** as it can develop into a complete human being. This is what happens when identical twins (or triplets or even quadruplets) form – such twinning can occur up to 14 days after conception. (These cells are called 'totipotent' as they have the 'potential' to develop into a 'total' individual.)

Q3.14 In most animals, fertilisation is followed by a series of rapid cell divisions. In these first few divisions, the embryo divides without growing in size. Smaller cells are produced with each successive division. Large reserves of nutrients and extra cell contents that eggs accumulate allow the zygote to divide rapidly soon after fertilisation. Which phases of the cell cycle will be shortened to achieve this rapid division?

By five days after conception, a hollow ball of cells called the **blastocyst** has formed (Figure 3.28). The outer blastocyst cell layer goes on to form the **placenta**. The inner cell mass, of 50 or so cells, goes on to form the tissues of the developing embryo. These 50 cells are known as **pluripotent embryonic stem cells**. Each of these cells can potentially give rise to most cell types, though they cannot each give rise to all the 216 different cell types that make up an adult human body. (The 'pluri' of 'pluripotent' is like the word 'plural'. It means that the cell can give rise to many cell types but, unlike the earlier totipotent cells, not all of them.)

Figure 3.28 Five days or so after conception, a human zygote divides to form a hollow ball of cells. Between 6 and 12 days after fertilisation the embryo implants in the wall of the uterus. Magnification ×250.

Cells become more differentiated

As the embryo develops into a multicellular body, the cells from which it is made become increasingly **differentiated**. Most of them lose the capacity to develop into a wide range of cells. Instead they become increasingly specialised, functioning as a red blood cell, one of the cell types in bone, a plant xylem vessel or whatever (Figure 3.34, page 127). However, even in adults some cells retain a certain capacity to give rise to a variety of different cell types. These cells are known as

multipotent stem cells. For example, neural stem cells can develop into the various types of cell found in the nervous system, white blood stem cells, located in bone marrow, can develop into red blood cells, platelets and the various sorts of white blood cells (macrophages, lymphocytes and so on).

The presence of stem cells in an adult is not unique to humans, or even to mammals. For example, in the body walls of cnidarians, such as sea anemones and jellyfish, there are cells that on first sight appear to have no function. They do not help to defend the animal, feed it or protect it. They are just small, unspecialised cells.

Cnidarians have specialised cells that they use for paralysing their prey. These specialised cells work rather like harpoons and can only be fired once. They need to be constantly replaced and this is the role of the unspecialised cells, which divide and differentiate when more of the harpoon-like cells are needed.

Figure 3.29 Clones of the house plant *Streptocarpus*, common name Cape primrose, are produced by planting leaf segments in compost.

Unlike animal cells, most plant cells remain totipotent throughout the life of the plant. Differentiation is irreversible in animal cells but many differentiated plant cells can dedifferentiate and then develop into a complete new plant. Gardeners have long known of this ability to regenerate identical whole plant clones from root, stem or leaf cuttings (Figure 3.29).

Totipotency of plant cells allows plants to be reproduced using plant tissue culture. Small pieces of a plant, known as explants, are surface sterilised and then placed on a solid agar medium with nutrients and growth regulators. The cells divide to form a mass of undifferentiated cells known as a callus. By altering the growth regulators in the medium, cells of the callus can be made to differentiate to form small groups of cells that are very similar to plant embryos. These embryos develop into complete plants that are genetically identical clones. Plant tissue culture allows commercial growers to produce large numbers of genetically identical plants rapidly (Figure 3.30). Tissue culture is also important in plant biology research, plant breeding, genetic modification of plants and in the conservation of endangered plants.

Weblink

To find out how plant tissue culture and micropropagation are being used at Kew in the conservation of endangered plants, visit the Royal Botanic Gardens website.

Activity

In **Activity 3.11** you can demonstrate totipotency of plant cells using tissue culture techniques. **A3.11S**

Checkpoint

3.5 Write a clear definition for each of the following terms used to describe stem cells: totipotent, pluripotent and multipotent.

Figure 3.30 Cereal plants being grown in test tubes from tissue cultures. Tissue culture techniques allow large-scale production of genetically identical disease-free plants for horticultural and agricultural use.

Potential use of human stem cells in medicine

Stem cells offer great hope to medicine. They may one day produce universal human donor cells which would provide new cells, tissues or organs for treatment and repair by transplantation. Embryonic stem cells may be the most suitable type of stem cells for this sort of treatment. Their potential to develop into any cell type offers the greatest flexibility for development, unlike adult stem cells which are committed to developing only into certain cell types.

Pluripotent stem cells for research and medicine can be isolated from so-called 'spare embryos'. These are produced in fertility clinics that carry out *in vitro* fertilisation, where the ovum is fertilised outside the body. Women undergoing this treatment are given drugs to make them superovulate, producing more eggs than are needed for infertility treatment. Many of the embryos end up being placed in women's wombs in the hope of enabling infertile couples to have children, but any additional embryos could be a source of embryonic stem cells. The use of embryos in this way is controversial – see the section in Topic 2 about ethical concerns (section 2.8).

In the laboratory, embryos being used for research are allowed to grow to form blastocysts. At this point, the embryos are cultured for a further period of time to see if stem cells are formed. The stem cells are isolated from each embryo and the rest of the embryo (which could not develop further) is discarded. The stem cells are then cultured and used in research. In the future they might be used to develop tissues that could be used for transplantation (Figure 3.31).

One problem with this approach is that even if scientists manage to get the stem cells to develop into the right sort of tissue, the tissue might end up being rejected by the immune system of the person given the transplant.

Figure 3.31 Stem cells are being isolated from embryos for research and can be grown in culture with the potential to produce tissues or organs for transplantation. Researchers have successfully grown tissues such as tendons, cartilage, and heart value tissue from stem cells. Technically, though, this is difficult and the work is still very much at the research stage.

There are several ways in which it is hoped to get round this problem of transplant rejection. One way would be to use tissue typing. This practice has routinely been used for decades when finding a suitable donor for a blood transfusion. Tissue typing for the transplantation of whole organs is more complicated than for blood transfusions, but the principles are the same. If whole organs are one day grown from stem cells – and this is a very long way off – it has been estimated that perhaps only 20 stem cell lines would be needed to provide transplants for 90% of the UK population.

A second approach to the problem of rejection would be to use drugs that prevent the recipient from rejecting any transplanted organ. Again, this approach has been around for many years and is widely used with conventional organ transplants.

A third and novel approach has been proposed and is generally referred to as **therapeutic cloning**, though the use of this term is regretted by many of the scientists working in the area. In therapeutic cloning the patient needing a transplant would have one of their diploid cells removed – this could simply be a cell from the base of a hair or any other suitable tissue. This cell, or its nucleus, would then be fused with an ovum from which the haploid nucleus had been removed (Figure 3.32). The result would be a diploid cell rather like a zygote. This process is known as somatic cell nuclear transfer (a somatic cell is any diploid body cell).

This cell could then be stimulated to divide by mitosis in the same way as the cell which gave rise to Dolly the cloned sheep (see Figure 3.36, page 129). If all went to plan, after about five days a blastocyst would develop. Stem cells could then be isolated from this, and encouraged to develop into tissues. This procedure results in cell lines, and perhaps eventually organs for transplantation, which are genetically identical to the patient from whom the original diploid cell was taken.

In 2004, the UK Human Fertilisation and Embryology Authority (HFEA) gave permission for the use of therapeutic cloning in research. Somatic cell nuclear transfer is used to produce blastocysts from which human stem cells are extracted. These stem cells are used for research into diseases such as Parkinson's disease and motor neurone disease. The cloned embryos created have to be destroyed at 14 days old; it is against the law for cloned embryos to be used for reproductive purposes, even though this may technically be possible. The human ova come from consenting IVF patients but such eggs are in limited supply.

Figure 3.32 The ovum is held in place with a pipette. Its genetic material is removed and replaced with an adult cell nucleus.

In 2007, the HFEA decided to allow research using animal-human embryos. An animal ovum (e.g. from a cow) can be used rather than a human ovum so research is no longer limited by the supply of human ova. The nucleus of the animal egg is removed, and a human nucleus is fused with the egg cell; the resulting cloned embryos are a source of stem cells for research.

As it is possible to maintain embryonic stem cells in tissue culture (Figure 3.31), they are being used for research into both human development and disease. For example, they can be used to explore how genes trigger the onset of development, such as organ formation. Stem cell research may help us to understand how cancer cells develop and how certain birth defects occur. Stem cells could also be used to provide a source of normal human cells of virtually any tissue type for use in screening new drugs.

Ethical concerns about the use of stem cells

Just about everyone agrees that there are no ethical objections to using multipotent stem cells derived from adults, for example from adult bone marrow. The problem is that most scientists believe that these stem cells are likely to be less valuable for research and for the development of new treatments than the pluripotent stem cells, which can only be derived from human embryos. Even if adult stem cells are used for new research, they may be of little use unless a better understanding is gained of how they specialise. This understanding may only come from embryonic stem cells.

Different people see the status of the human embryo very differently. As the official UK government committee set up to report on human stem cell research – with the Chief Medical Officer, Professor Liam Donaldson, in the chair – said in the year 2000:

> *A significant body of opinion holds that, as a moral principle, the use of any embryo for research purposes is unethical and unacceptable on the grounds that an embryo should be accorded full human status from the moment of its creation. At the other end of the spectrum, some argue that the embryo requires and deserves no particular moral attention whatsoever. Others accept the special status of an embryo as a potential human being, yet argue that the respect due to the embryo increases as it develops and that this respect, in the early stages in particular, may properly be weighed against the potential benefits arising from the proposed research.*

The role of the regulatory authorities

Much scientific work can simply be applied usefully without the need for any regulatory authorities to get involved. However, there are times when new advances in society raise various ethical or other concerns and in these cases, regulation may be needed.

In the UK, the ultimate regulatory authority is parliament. New laws are passed by majorities in both the House of Commons and the House of Lords and then the wording of these laws is interpreted by the courts.

In matters of science, those responsible for making new laws can receive advice from a number of sources. For example, there are 'Select Committees' in both the Lords and the Commons. These are made up of peers of the realm (in the Lords) and MPs (in the Commons), but they are advised by specialist advisors who have particular expertise in the subject in question.

There are also various Advisory Committees. For example, the Advisory Committee on Novel Foods and Processes advises the government as to whether new foods (including foods made from genetically modified ingredients) should be permitted. Its members are mainly independent scientists but also include an ethicist.

Scientists make an objective interpretation of the scientific evidence. This scientific knowledge, and public consultation, help the regulatory authorities make decisions about the applications and implications of the science in the context of society.

In the UK, the Human Fertilisation and Embryology Authority (HFEA) regulates research on human embryos. Until 2001, UK law only allowed the use of human embryos where the HFEA considered their use to be necessary or desirable:

- to promote advances in the treatment of infertility
- to increase knowledge about the causes of congenital disease
- to increase knowledge about the causes of miscarriage
- to develop more effective methods of contraception
- to develop methods for detecting gene or chromosome abnormalities in embryos before implantation.

On 22 January 2001, peers in the UK House of Lords voted by 212 to 92 votes to extend the purposes for which research on human embryos is allowed. (A vote in the House of Commons had already gone the same way in December 2000 by 366 to 174 votes.) So-called 'spare' embryos from *in vitro* fertilisation treatment can now be used as a source of embryonic stem cells for the purpose of research into serious disease. On 27 February 2002 this was extended to include fundamental research necessary to understand differentiation and dedifferentiation of cells.

These 2001 and 2002 extensions have led the HFEA to permit new areas of work, including its 2007 decision to allow research using animal-human embryos. However, this whole area of work is very controversial and it is possible that those opposed to such work may attempt to take the HFEA to court. Ultimately it is parliament that lays down the laws and it is the courts that decide whether or not laws have been broken.

Activity

In **Activity 3.12** you can read some quotes from those who are for and against the use of stem cells and decide what your own position is on the issue. **A3.12S**

3.3 How development is controlled

In Figure 3.33 you can see an early stage in the development of a mouse. As cells divide after fertilisation they become specialised for a small number of functions (Figure 3.34). As you probably already know, specialised cells often work together as tissues in organs.

Extension

It is possible to watch the growth and development of zebrafish through a light microscope because they are transparent. In **Extension 3.4** you can watch their development and find out how these fish are being used to study development and genetics. You can also see the early stages in human development. **X3.04S**

Figure 3.33 A mouse during an early stage in its development.

Figure 3.34 Although these specialised cells look different to one another, they have all been produced by mitosis from a common ancestor so we know they must all contain a complete complement of chromosomes with the same genetic information.

We now take it for granted that the nucleus has a role in controlling the development of the individual cell and the whole multicellular organism's phenotype. This was first shown in classic experiments using giant algal cells, completed in 1943 by a Danish biologist named Joachim Hammerling (Figure 3.35).

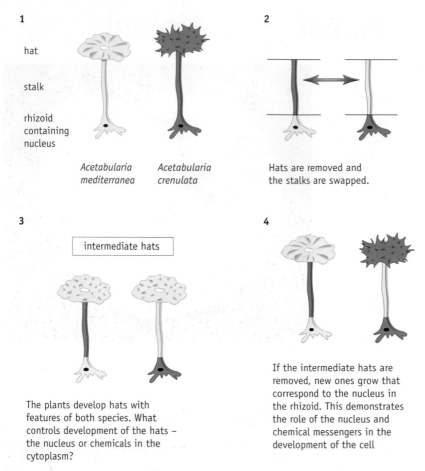

Figure 3.35 Is it the nucleus or the cytoplasm that controls development in *Acetabularia mediterranea* and *A. crenulata*?

Activity

Using the simulation in **Activity 3.13** you can carry out a simulation of Hammerling's experiments, demonstrating the role of the nucleus in controlling development of giant algal cells. **A3.13S**

Q3.15 In 1997, Ian Wilmut and his colleagues at the Roslin Institute in Scotland successfully cloned an adult sheep. They transplanted the nucleus from the mammary gland cell of one adult sheep into another sheep's unfertilised ovum from which the nucleus had been removed. The diploid cell that formed developed into an embryo which was implanted into another adult sheep (Figure 3.36). The surrogate mother gave birth to a lamb who soon became world famous, 'Dolly'.

How did the creation of Dolly support the idea that all the genetic information for making a complete organism is present in every single cell, including those in an adult with specialised functions?

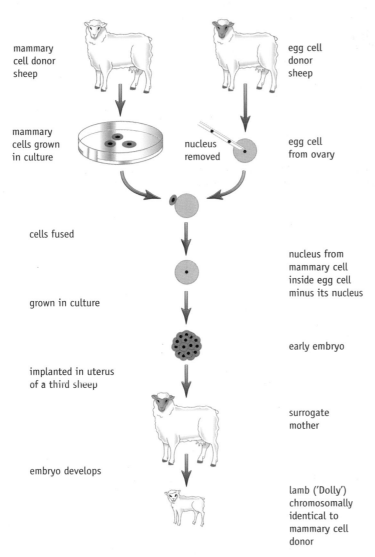

mammary cell donor sheep

egg cell donor sheep

mammary cells grown in culture

nucleus removed

egg cell from ovary

cells fused

nucleus from mammary cell inside egg cell minus its nucleus

grown in culture

early embryo

implanted in uterus of a third sheep

surrogate mother

embryo develops

lamb ('Dolly') chromosomally identical to mammary cell donor

Figure 3.36 Dolly's genes were identical to those of the mammary cell donor sheep.

? Did you know?

Problems with cloning

Animal cloning is not yet as easy as you might think from the description in Q3.15. In fact most developmental biologists (including Ian Wilmut) feel that the process is not safe to try on humans. Many attempts to produce live-born animals by cloning have been unsuccessful and a high proportion of the animals produced by cloning have health problems. Dolly herself was the sole success from 277 attempts at cloning and this low success rate is still typical. Even Dolly may not have been completely 'normal' as she developed arthritis at a fairly young age for a sheep and had to be put down.

The reasons why cloning mammals from adult nuclei has such a low success rate are not yet fully understood, and there are probably several different contributing factors. One of these is that the DNA in an adult cell nucleus has been programmed into a particular type of cell (e.g. skin cell). When transferred to an ovum, the nucleus may not be able to reprogram its DNA quickly enough to be able to switch on all the different genes required for normal development.

Different genes are expressed

As the embryo develops, cells differentiate: they become specialised for one function or a group of functions. The structure and function of each cell type is dependent on the proteins that it synthesises. Salivary gland cells make salivary amylase, an enzyme for digesting starch, while red blood cells contain the protein haemoglobin, an oxygen-carrying pigment. Amylase and haemoglobin are both proteins coded for by the cells' genes. How is it that these cells, and only these cells, produce these proteins although every cell contains the complete genome including the codes for both amylase and haemoglobin? Different specialised cells must only be expressing some of their genes.

Work by Igor Dawid and Thomas Sargent demonstrated that different genes are expressed in different cells. They extracted mRNA from differentiated and undifferentiated frog cells (Figure 3.37). Complementary DNA (cDNA) strands were produced for all the mRNA in the differentiated cells using an enzyme called reverse transcriptase, which, as its name suggests, reverses transcription, making DNA from mRNA. These cDNA strands were mixed with the mRNA from the undifferentiated cell. Complementary strands of cDNA and mRNA combined to produce double strand hybrids. When these hybrids were separated out, there remained a range of cDNA strands that had not been hybridised; the two cells were expressing some of the same genes but also some different genes.

Figure 3.37 How Dawid and Sargent demonstrated differential gene expression in the cells of developing frogs.

Switching genes on

Cells become specialised because only some genes are switched on and produce active mRNA which is translated into proteins within the cell.

Switching on β-galactosidase synthesis

The French geneticists Jacob and Monod were the first to propose a theory for the control of gene expression in the early 1960s. They studied the control of genes in the prokaryote *Escherichia coli*. These bacteria only produce the enzyme **β-galactosidase** to break down the carbohydrate lactose when it is present in the surrounding medium. This enzyme converts the disaccharide lactose to the monosaccharides glucose and galactose.

When lactose is not present in the environment, a lactose repressor molecule binds to the DNA and prevents the transcription of the β-galactosidase gene. Figure 3.38 shows how the lactose repressor stops the β-galactosidase gene being expressed. If lactose is present in the environment, the repressor molecule is prevented from binding to the DNA, and the β-galactosidase gene is expressed. mRNA coding for β-galactosidase is transcribed and translation of this mRNA produces the enzyme.

Activity

In **Activity 3.14** you can see how a gene is switched on by the presence of a substrate. **A3.14S**

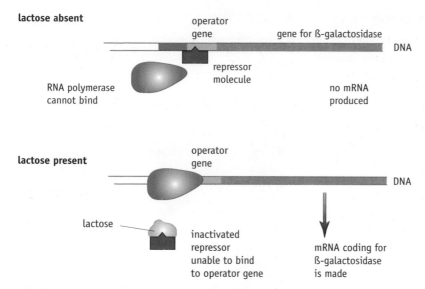

Figure 3.38 Control of the β-galactosidase gene in the bacterium *Escherichia coli*. The RNA polymerase must bind for transcription to start.

What switches an individual gene on or off in eukaryotes?

Genes in uncoiled, accessible regions of the eukaryote DNA can be transcribed into messenger RNA. The enzyme RNA polymerase binds to a section of the DNA adjacent to the gene to be transcribed. This section is known as the promoter region. Only when the enzyme has attached to the DNA will transcription proceed. This process is outlined in Figure 3.39. The gene remains switched off until the enzyme attaches to the promoter region successfully. The attachment of a regulator protein is usually also required to start transcription.

Figure 3.39 DNA transcription will start only when the RNA polymerase has attached to the promoter region of the gene that is to be transcribed.

Transcription of a gene can be prevented by protein repressor molecules attaching to the DNA of the promoter region, blocking the attachment site. In addition, protein repressor molecules can attach to the regulator proteins themselves, preventing them from attaching. In either case the gene is switched off; it is not transcribed within this cell.

Extension

In **Extension 3.5** read about how one X chromosome is inactivated in every female mammalian cell. **X3.05S**

When gene expression goes wrong

The importance of differential gene expression in producing specialised cells is highlighted when things go wrong. Here is an example.

Growing bones in the wrong places

Diana is a baby whose big toes are too short and are curved inwards towards her other toes. When her parents took her to a specialist they were told that this can be the first sign of a rare inherited disease called FOP (fibrodysplasia ossificans progressiva). FOP is characterised by the growth of bones in odd places, such as within muscle and connective tissue (Figure 3.40). By early adulthood, the abnormal bone growth often leads to the 'freezing' of all major joints of the backbone and limbs so they cannot move. Any local trauma, even injection of medicines into muscle, can cause bone formation in people with FOP.

What causes FOP?

FOP is an inherited condition caused by a gene mutation. Bone cells are normally only produced in growing limbs and in other places where the skeleton develops. Here, genes are expressed that produce all the proteins needed to become a specialised bone cell. In people with FOP, one of these genes is not switched off in white blood cells. When tissue is damaged, for example when having an injection, the body's white blood cells move to the site of the damage. The white cells produce the protein, which diffuses into the surrounding muscle cells. The protein causes the muscle cells to start expressing other genes that turn them into bone cells.

How are cells organised into tissues?

Specialised cells can group themselves into clusters, working together as a tissue. Cells have specific recognition proteins, also known as adhesion molecules, on their cell surface membranes. Adhesion molecules help cells to recognise other cells like themselves and stick to them. A small part of each recognition protein is embedded in the cell surface membrane; a larger part extends from the membrane. This exposed section binds to complementary proteins on the adjacent cell (Figure 3.41). The particular recognition proteins synthesised by a cell determine which cells it can and cannot attach to. If cells from different tissues are separated and then mixed together they reform into the tissues as the recognition proteins bind.

In the human embryo, cells begin to form tissues as they start their specialised functions. At this point the genes coding for the recognition proteins are switched on. Reorganisation of tissues during development occurs through changes in the type of recognition proteins produced by the cells.

Figure 3.40 Bones form in the wrong places in people with the rare inherited disease called FOP. By their early 20s, most patients are confined to wheelchairs. Starvation used to result from the freezing of the jaw, and pneumonia can occur due to the lungs or diaphragm becoming fixed to the chest wall. Surgery can make the condition worse by causing more bone growth. There are no proven treatments that prevent progression of the disease.

cytoplasm

Cell surface membrane

Complementary recognition proteins bind

tissue fluid

Cell surface membrane

cytoplasm

Figure 3.41 Recognition proteins synthesised by cells determine which cells work together in tissues and organs.

In tissues, cells also interact with the extracellular matrix, a network of molecules secreted by cells. In some tissues the extracellular matrix is a major component of the tissue, for example in cartilage.

Table 3.1 summarises the levels of organisation within organisms.

cell	In multicellular organisms cells are specialised for a particular function. For example, muscle cells and epithelial cells.
tissue	A group of specialised cells working together to carry out one function. For example, muscle cells combining to form muscle tissue, and epithelial cells forming epithelial tissue.
organ	A group of tissues working together to carry out one function. For example, muscle, nerve and epithelium work together in the heart.
organ systems	A group of organs working together to carry out a particular function. For example, the circulatory system.

Table 3.1 Summary of the levels of organisation within organisms.

Q3.16 State **a** the level of organisation and **b** the function performed by each of the following examples:

- xylem vessel in a flowering plant
- leaf
- kidney
- neurone.

Gene expression and development

The precise sequence of transcription and translation of genes determines the sequence of changes during development. By studying the embryonic development of model animals and plants, researchers have shown how genes determine the structures produced during development.

Master genes

In fruit flies, *Drosophila*, once the main body segments have been determined, the cells in each segment become specialised for the appropriate structures in that segment, for example, legs, antennae and wings. Master genes control the development of each segment. These genes were discovered by looking at mutations which cause the development of the wrong appendage for that segment (Figure 3.42). The master genes produce mRNA which is translated into signal proteins. These proteins switch on the genes responsible for producing the proteins needed for specialisation of cells in each segment. You will study how these signal proteins switch on genes in Topic 7.

Figure 3.42 **A** In this mutant fruit fly, legs have developed where the antennae should be. **B** The head of a normal fruit fly. Magnification ×90.

The ABC of flowering plants

When a plant starts to flower, cells in a meristem become specialised to form the organs that make up the flower. Flowering of *Arabidopsis thaliana* has been used to model what is happening when plants change from vegetative growth to reproductive development. Most hermaphrodite flowers, those that have male and female structures, contain four sets of floral organs: sepals, petals, male stamens and female carpels. These are arranged in concentric whorls (Figures 3.43 and 3.44).

The expression of genes in cells across the meristem determines which structures will form. Three genes determine which type of specialised organ will be produced in each area of the meristem (Figure 3.44). Where only gene A is expressed sepals form, and where only gene C is expressed carpels form. Petals form where A and B genes are expressed, and stamens form where B and C genes are expressed. As with the master genes in fruit flies, these genes produce mRNA that codes for signal proteins which switch on appropriate genes. Synthesis of proteins coded for by these genes results in development of specialised cells in each floral organ.

Figure 3.43 *Arabidopsis thaliana* (wall cress), a member of the mustard family, is used extensively in genetic and molecular research.

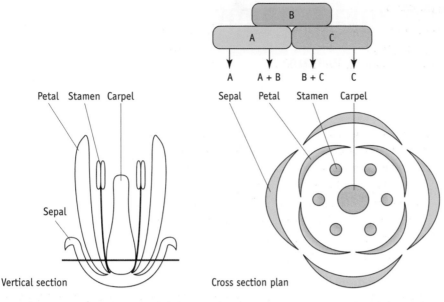

Figure 3.44 The role of genes in floral development. The ABC genes active in the formation of the flower are shown at the top, with their corresponding structures highlighted on the cross-section plan of the flower below.

Q3.17 The role of the three master genes that control flower development was determined by looking at plants with mutations. One mutation causes a plant to only produce genes A and B.

a Name the floral structures that would form within the flowers of this plant.

b Describe how these would be arranged in the flower (a diagram would be useful).

Activity

In **Activity 3.15** you construct flowers using the ABC model. **A3.15S**

Checkpoint ✔

3.6 Write a short paragraph which summarises how cells become specialised for their different structures and functions.

? Did you know?

Programmed cell death (apoptosis)

It is an amazing fact that all mammal cells have a 'self-destruct' programme. During development an animal must lose some cells by programmed cell death, called apoptosis. There is a small group of cell 'suicide' genes, and when they are expressed this causes the nucleus and cytoplasm to fragment.

Genes that prevent death are expressed in most cells, and during development cells have to switch on their 'suicide' genes in order to die. This explains why cell death does not happen throughout the embryo. Programmed cell death occurs in particular places, such as between where the fingers will be. In the embryo brain many millions of extra cells are produced, at least half of which die by controlled cell death.

3.4 Genes and environment

Is it all in the genes? Nature and nurture

The characteristics of an organism, such as its height, shape, blood group or sex, are known as its **phenotype**. Differences in phenotype between the members of a population are caused by differences in:

- genetic make-up or **genotype**
- the environment in which an individual develops.

Some characteristics are controlled almost completely by the organism's genotype, with the environment having little or no effect. For example, a person's blood group (group A, B, AB or O) is controlled by the genes they inherit. The genes code for protein antigens on the surface of their red blood cells. A person's blood group is not affected by the environment in which they develop. Such characteristics are controlled by genes at a single locus and show **discontinuous variation**. They have phenotypes that fall into discrete groups with no overlap (Figure 3.45A). In garden peas, unlike humans, height also shows discontinuous variation.

Characteristics that are affected by both genotype and environment often show **continuous variation**. Human height is a characteristic showing continuous variation. This means that a person can be any height within the human range. The most common height will be somewhere mid-way between the extremes of the range. If a graph is drawn showing the frequency distribution of the different height categories it will be bell-shaped (Figure 3.45B). Characteristics that show continuous variation are controlled by genes at many loci, known as polygenic inheritance, and also by the environment, either directly or by influencing gene expression.

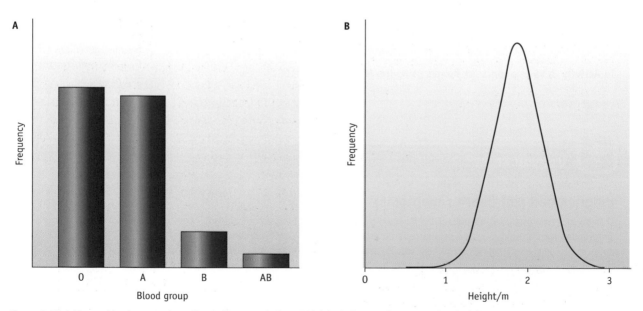

Figure 3.45 A Human blood groups show discontinuous variation. **B** Height in humans is an example of continuous variation.

Key biological principle: Polygenic inheritance

In monohybrid inheritance, each locus is responsible for a different heritable feature. For example, one gene might be coding for the colour of a flower with another gene coding for the shape of the petals. However, inheritance of most characteristics does not follow simple Mendelian rules of inheritance, but involves interaction of alleles at many loci. When a number of genes are involved in the inheritance of a characteristic, rather than just one, we call the pattern of inheritance **polygenic**.

In many human characteristics that show polygenic inheritance, such as height and skin colour, the alleles clearly have additive effects. With conditions such as diabetes, coronary heart disease, Alzheimer's disease, schizophrenia and some cancers, several genes may confer a *susceptibility* to the condition, with environmental factors also contributing. Two people who inherit the same susceptibility may not both develop the illness; it will depend on environmental factors such as diet, exposure to toxins and stress. Such factors act as triggers to bring about the symptoms of disease. Conditions where several genetic and one or more environmental factors are involved are said to be **multifactorial.**

The degree of similarity between identical twins is a measure of the influence of the genes on that characteristic. 99% of identical twins have the same eye colour, and 95% the same fingerprint ridge count. Where the environment has a greater effect, the similarity falls. One study found that if you are an identical twin and you have Alzheimer's disease, your twin sibling (brother or sister) has a 40% chance of having Alzheimer's disease. However, if you have Alzheimer's disease as a non-identical twin, your sibling has only a 10% chance of having Alzheimer's disease. This suggests that there is a significant but not inescapable genetic basis to Alzheimer's disease. Both genes and the environment influence the development of the disease.

How polygenic inheritance works

In any introductory course on genetics, it is common for eye colour to be used as an example of monohybrid inheritance – a single locus with brown dominant to blue. This is not entirely the case. Eye colour is an example of polygenic inheritance; alleles at several loci control eye colour. Eye colour ranges from blue to brown, depending on the amount of pigment in the iris. The pigment absorbs light so brown eyes appear dark. Blue eyes have little pigment, so light reflects off the iris.

Let us say three loci are involved in the inheritance of this characteristic, each with alleles **B** and **b**. **B** adds pigment to the iris and **b** does not. If all three loci were homozygous for the allele **B**, the person's genotype would be **BB BB BB**. The additive effect would produce a dark brown iris, whereas **bb bb bb** would add no pigment

to the iris, making it pale blue. A range of possible genotypes and phenotypes are possible between these two extremes, according to how many alleles add brown pigment, as shown below.

Number of alleles adding brown pigment	Example of genotype	Eye colour
6	BB BB BB	very dark brown
5	BB BB Bb	dark brown
4	BB BB bb	medium brown
3	BB Bb bb	light brown
2	BB bb bb	deep blue
1	Bb bb bb	medium blue
0	bb bb bb	pale blue

Table 3.2 Eye colour is a polygenic characteristic

The greater the number of loci involved, the greater the number of possible shades.

If a pale blue-eyed woman has children with a very dark brown-eyed man they will have light brown-eyed children as shown below.

	Mother	Father
Parental phenotypes	Pale blue	Very dark brown
Parental genotypes	**bb bb bb**	**BB BB BB**
Gametes	**bbb**	**BBB**
Offspring genotypes	**Bb Bb Bb**	
Offspring phenotype	Light brown	

Q3.18 A deep blue-eyed woman (**Bb bb bB**) has a child with light brown eyes. Her medium blue-eyed partner (**Bb bb bb**) suspects that he is not the father. Copy and complete the Punnett square below and describe how you would use it to explain to him that he could be the father.

	Mother	Father
Parental phenotypes	Deep blue	Medium blue
Parental genotypes	**Bb bb bB**	**Bb bb bb**
Possible gametes	**Bbb, BbB, bbb, bbB**	**Bbb, bbb**

		Gametes from the father	
		Bbb	**bbb**
Gametes from the mother			

CONTINUED ▶

Height, weight and skin pigmentation all involve polygenic inheritance. Each allele has a small effect on the characteristic, and the effects of several alleles combine to produce the phenotype of an individual. Suppose that only two genes were involved in the determination of height. The homozygous recessive genotype, **aabb**, might give a height of 20 cm above a baseline of 140 cm for adult women and 150 cm for adult men. In other words the recessive alleles (**a** and **b**) each contribute 5 cm to the height. The dominant alleles, **A** and **B**, each contribute 10 cm to the height – so the homozygous dominant **AABB** would give a height of 40 cm above our baselines. **AaBb** would add 30 cm to the height.

If two heterozygotes were crossed there would be a range of phenotypes as shown below.

Height above baseline/cm	20	25	30	35	40
Number of offspring with that height	1	4	6	4	1

Table 3.3 Range of phenotypes from crossing two **AaBb** heterozygotes.

Q3.19   **a**   Sketch what a graph of the data from Table 3.3 would look like.

b   What can you conclude about height from this graph?

c   For the cross between two heterozygotes, i state the parental phenotypes and possible gamete genotypes, and ii draw up a Punnett square to show the possible offspring genotypes and phenotypes.

If instead of two loci there were several, the number of height phenotypes would increase. The greater the number of loci, the greater the number of height classes, and the smaller the differences between classes. If plotted, this would produce a smooth bell-shaped curve typical of continuous variation (see Figure 3.45B).

Traits such as height and skin colour are not just the result of an individual's genotype. Their environment also has an influence. For example, diet has an effect on a person's height. A poor diet may prevent a person reaching their full height as predicted by their genotype. Their genotype gives their *potential* height. Their genotype in combination with their environment determines their *actual* height. The result is that there are no clear height classes, but a continuous variation in height between the extremes of the range.

Activity

In **Activity 3.16** discover how many genes can affect a single characteristic. **A3.16S**

Gene and environment interactions

There are countless examples of genes and environment interacting to produce an organism's phenotype. Some are very familiar, such as skin and hair colour; others are less so. We will take a detailed look at four examples:

- height
- hair colour
- MAOA
- causes of cancer.

Height

Have you noticed that each human generation seems to be a bit taller than the last (Figure 3.46)? The average human height in industrialised countries has risen each generation over the past 150 years. People in the UK now average about 8 cm taller than they did in 1850. How can this be explained?

Before reading on, list as many reasons as you can for the increase in human height over recent generations.

The following are some possible answers. You may have thought of others.

1 There is some evidence that taller men have more children, which would result in a gradual change in the genetic make-up of the population.

2 Greater movements of people have resulted in less inbreeding, leading to taller people.

3 Better nutrition, especially increased protein, has resulted in greater growth of children.

4 Improved health, with a reduction in infectious diseases, has occurred through improved sanitation, clean water supplies, vaccination and antibiotics.

5 The end of child labour has allowed more energy to go into growth.

6 Better heating of houses and better quality clothing reduces the amount of energy needed to heat the body, so again more energy can go into growth.

Q3.20 Which of these reasons for change in height are due to the environment and which are the result of genotype?

Figure 3.46 People are getting taller.

Activity

Take part in **Activity 3.17** to see if the difference in heights over the years can be observed. **A3.17S**

It is widely accepted that a person's height is determined by an interaction between the effects of their genes for height and environmental influences, such as diet. A person may have genes for being tall, but not achieve his or her full potential height because of malnutrition. We do not know for certain which of the possible reasons for change in height is the most important. However, most people think that better diet is the most significant factor.

Extension

In **Extension 3.6** find out how both genotype and environment influence sex determination. **X3.06S**

Hair colour

Differences in hair colour are largely genetically determined, due to variation in the amount and type of pigment the hair contains. But the environment can influence hair colour in some cases.

Making melanin

The dark pigment in skin and hair is called **melanin**. Melanin is made in special cells, melanocytes, found in the skin and at the root of the hair in the follicle. These are activated by **melanocyte-stimulating hormone** (**MSH**). There are receptors for MSH on the surface of the melanocyte cells. The melanocytes place melanin into organelles called melanosomes. The melanosomes are transferred to nearby skin and hair cells where they collect around the nucleus, protecting the DNA from harmful UV light. People with more receptors have darker skin and hair; they have more protection against sunburn. Ultraviolet (UV) light increases the amount of MSH and also of MSH receptors, making the melanocytes more active (Figure 3.47) and causing the skin to darken. Hair does not appear darker because although more melanin is produced, UV light causes chemical and physical changes to melanin and other proteins in hair cells. Hair lightens due to destruction of the melanin by UV light.

A Low exposure to UV light

B Increased exposure to UV light

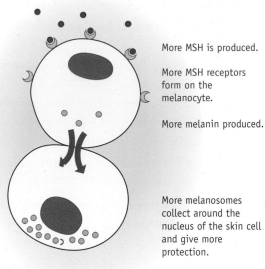

melanocyte-stimulating hormone (MSH)

MSH receptor

melanocyte cell

Melanin is made and put into melanosomes.

skin cell

Melanosomes are transferred to skin cell.

nucleus

Melanosomes collect around the nucleus, protecting it from damage by UV light.

More MSH is produced.

More MSH receptors form on the melanocyte.

More melanin produced.

More melanosomes collect around the nucleus of the skin cell and give more protection.

Figure 3.47 The formation of melanin in melanocytes.

Seasonal colour change

Arctic foxes have brown fur in summer and white fur in winter (Figure 3.48). They must have the genes for making brown fur (which contains melanin) all the time, so how can white fur be made?

The white winter coat is actually grown during the summer. It grows under the brown summer coat and is revealed when the summer coat is moulted in autumn. The foxes produce fewer MSH receptors in the summer. Without these receptors, MSH has no effect and no melanin is made in the hair follicles.

Q3.21 Explain why MSH has no effect on the developing hairs in summer, meaning that the coat grows white.

Q3.22 Explain why it is surprising that white fur is able to grow in the summer.

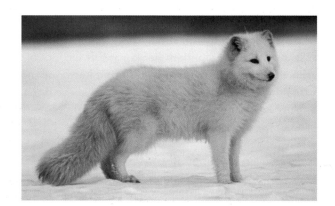

Figure 3.48 How does the arctic fox change colour?

White with dark tips

To make melanin, animals use an enzyme called **tyrosinase**. Tyrosinase catalyses the first step along a chemical pathway, changing the amino acid tyrosine into melanin. Some animals, such as Himalayan rabbits and Siamese cats, have mutant alleles for tyrosinase. The enzyme is made but it is unstable and is inactivated at normal body temperature. However, the tips of their tail, paws and ears are much darker than the rest of their bodies (Figure 3.49).

Figure 3.49 Only the ears, paws and tail of this Himalayan rabbit are dark.

Q3.23 Suggest how the environment and genotype are interacting to produce the distinct colouring at the tail, paws and ear tips in Figure 3.49.

Q3.24 There are rare cases in humans in which hair in the armpits is white, but hair on places such as the legs is dark. Suggest how this happens.

MAOA

Monoamine oxidase A is an enzyme that catalyses the breakdown of a neurotransmitter in the brain involved in the regulation of behaviour, including the response to stress. It was found that some individuals have a rare mutation in the MAOA gene and produce no enzyme. They exhibit aggressive and sometimes violent behaviour. Genetically modified mice which lack the MAOA gene are also aggressive. These observations led scientists to suggest there might be a connection between the gene and violent behaviour but studies did not show a clear link.

Researchers at King's College in London revealed an interaction between MAOA genotype, mistreatment in childhood and antisocial behaviour in later life. They used the results of a study that had followed the health and social development of over 1000 children in Dunedin, New Zealand, for over 20 years. There are two alleles for the MAOA gene. One results in the production of high levels of the enzyme, the other is linked with low level production of the enzyme. The researchers focused on the boys in the study because the MAOA gene occurs on the X chromosome so males have only one allele.

Childhood maltreatment was associated with more antisocial behaviour as adults. No direct link was found between MAOA levels and subsequent antisocial behaviour. However, maltreated children with high levels of MAOA were less likely to exhibit violent behaviour than maltreated children with low MAOA. 12% of the males had low MAOA and had been maltreated, but they committed 44% of the violent crimes recorded for the whole group. Further studies have supported this complex interaction between childhood environment and MAOA genotype.

Cancer

What causes cancer?

One in three people in the UK will suffer from cancer at some stage in their life and, at present, one in four people dies from the disease. If we could understand the way our genetic make-up and our environment combine to cause cancer, we would be on the way to finding means of prevention and cure.

Cancers occur when the rate of cell multiplication is faster than the rate of cell death. This causes the growth of a **tumour**, often in tissues with a high rate of mitosis, such as the lung, bowel, gut or bone marrow. Cancers are caused by damage to DNA. DNA is easily damaged by physical factors such as UV light or asbestos. It can also be damaged by chemicals, known as carcinogens, which may be in the environment or can be

> **Activity**
>
> **Activity 3.18** lets you look at how the relative contributions of genes and the environment are difficult to determine. **A3.18S**

produced by cell metabolism. Mutations can also occur when cells divide (see Topic 2, Section 2.5). If DNA is copied incorrectly in gamete formation, an inherited form of cancer can result.

Q3.25 Explain why cancers are more likely to occur in tissues with a high rate of mitosis.

Q3.26 Explain why damage to the DNA in an embryo can result in inherited cancer.

Telling cells what to do

Cells go through a fixed sequence of events during the cell cycle – G1, S, G2 and mitosis (M) – as shown in Section 3.2. The progression from one phase to the next is controlled. During each stage of the cycle, proteins are produced that stimulate the next stage in the cycle. Cells also produce proteins that stop the cell cycle, preventing progress from one stage to the next. These proteins activate or inhibit enzymes that initiate the reactions in the next stage of the cycle.

Cancer cells do not respond to the control mechanisms. Two types of gene have a role in control of the cell cycle and play a part in triggering cancer. These are:

- oncogenes
- tumour suppressor genes.

Oncogenes code for the proteins that stimulate the transition from one stage in the cell cycle to the next. Mutations in these genes can lead to the cell cycle being continually active. This may cause excessive cell division, resulting in a tumour.

Tumour suppressor genes produce suppressor proteins that stop the cycle. Mutations inactivating these genes mean there is no brake on the cell cycle. One example of a tumour suppressor protein is p53. This protein stops the cell cycle by inhibiting the enzymes at the G1/S transition, preventing the cell from copying its DNA. In cancer cells a lack of p53 means the cell cannot stop entry into the S phase. Such cells have lost the control of the cell cycle. Loss of tumour suppressor proteins has been linked to skin, colon, bladder and breast cancers.

There is a very complex network of signals and inhibitors that interact to control the cell cycle. There needs to be damage to more than one part of the cell control system for cancer to occur. This makes cancer very unlikely in any particular cell, but because the body contains so many cells dividing and changing over a lifetime, cancers will occasionally occur. Cancers are more likely in older people as they have accumulated more mutations.

Q3.27 In some people one of two alleles for the tumour suppressor protein p53 is damaged. The damaged allele is recessive to the normal allele. Explain why such people are more susceptible to environmentally induced cancer than people with two normal alleles for p53.

Inherited cancer

Most common cancers occur more frequently in close relatives of cancer patients, suggesting an inherited component. Many gene defects have been identified that predispose people to cancers including bowel cancer, ovarian cancer, prostate cancer, retinal cancer and some types of leukaemia.

For example, mutations in the gene BRCA1 predispose a person to breast cancer. The functioning BRCA1 gene produces a protein used to repair DNA. A child who inherits one defective BRCA1 allele may get cancer later in life if the other allele becomes

Extension

In **Extension 3.7** find out how enzymes control the cell cycle at checkpoints. **X3.07S** damaged in breast tissue cells. Having a single defective BRCA1 allele does not therefore mean that breast cancer is inevitable. It simply means that such individuals are more susceptible to cancer through environmental DNA damage. Women who inherit a single BRCA1 mutation have about a 60% chance of developing breast cancer by the age of 50 compared with only a 2% chance for those who inherit two normal BRCA1 alleles. The mutation confers a high risk but is relatively rare, accounting for only 5% of breast cancer cases.

Environment and cancer

You will have noticed that newspapers and magazines are full of suggestions for living a healthy life and reducing cancer risk. How useful are these suggestions?

Damage from the environment can be either chemical or physical. The greatest chemical risk of all is from smoking. Smoking increases the likelihood of many forms of cancer, especially lung cancer, through the action of carcinogens in tar. Tar lodges in the bronchi and causes damage to DNA in the surrounding epithelial cells.

Ultraviolet light (UV) physically damages DNA in skin cells. Sometimes, a mole which has been affected by UV light may start to grow bigger, and can develop into a tumour (Figure 3.50). If a tumour is not removed, cancer cells sometimes spread to other parts of the body, carried in the blood and lymphatic systems. New cancers may then form in other organs.

Figure 3.50 UV damage can cause cancer in skin cells.

Diet is also linked to cause and prevention of cancer, though the connections are not always clear. A diet with plenty of fresh fruit and vegetables provides antioxidants which destroy radicals (see Topic 1). Radicals are chemicals from the diet, from environmental factors such as smoke and UV, or produced by the cell's own metabolism, which contribute to ageing and cancer through DNA damage.

Several cancers are triggered by virus infection. For example, liver cancer can follow some types of hepatitis, and cervical cancer can follow infection by the papilloma (genital wart) virus. A virus's RNA may even contain an oncogene, which it has picked up from one of its hosts and then transfers to the cells it infects.

Q3.28 Explain why chemotherapy and radiotherapy are often unsuccessful in tumours where the cause of the cancer is damage to the protein p53.

? Did you know?

Combating cancer

One way of treating cancer is to use surgery to physically remove the tumour. Another is to destroy the cells in the tumours. In chemotherapy powerful chemicals are used to do this, and in radiotherapy X-rays or other radiation are directed at the tumour. The difficulty is to target the tumour cells without damaging nearby healthy tissues.

It used to be thought that chemotherapy and radiotherapy work by preferentially killing dividing cells in the tumour. It is now believed that these treatments actually work by inducing cells to carry out cell suicide. Chemotherapy and radiotherapy cause some DNA damage, but not enough to kill tumour cells. However, the DNA damage causes the release of the protein p53, so the cancer cells stop dividing.

Activity

Use **Activity 3.19** to check your notes using the topic summary provided. **A3.19S**

Topic test

Now that you have finished Topic 3, complete the end-of-topic test before starting Topic 4.

Biodiversity and natural resources

Why a topic called Biodiversity and natural resources?

The Brazil nut 'story'

What do the four organisms in Figure 4.1 have in common? You are correct if you suggest that they all live in the same habitat. They can all be found living in the Amazonian rainforest. Even better if you suggest that they are dependent on each other for their survival. The story of the Brazil nut tree is amazing. The tree grows to over 50 m in height; its fruit is the size of a grapefruit and weighs about two kilograms. Each fruit contains about 20 seeds – we know these as Brazil nuts. The fruits remain in the canopy for over 15 months before falling to the forest floor. On the ground there is only one animal with strong enough teeth to gnaw open the extremely hard fruit – a small rodent, the agouti. Once open, the agouti eats most of the Brazil nuts immediately, but buries some to be recovered later. If the Brazil nuts remain undisturbed by the agouti or other herbivores for over a year, their shells crack and they germinate. Brazil nut seed dispersal is completely dependent on this one animal.

Pollination of the Brazil nut tree is also specialised, relying on just one type of bee. Euglossine bees (also called orchid bees) are the trees' only pollinators; they are large enough and strong enough to get into the tightly hooded flowers. Female bees do most of the pollinating, using their long specialised tongues to reach the nectar.

The male bees visit orchid flowers that provide scent for them to attract female mates. The orchid benefits as the bee transports pollen to the next orchid flower visited. Without the orchids there would be no young bees, no adult bees, and no Brazil nuts. All four species are inextricably linked; if one became extinct, the others would not be able to survive.

And it is not just these species that depend on the Brazil nut tree. When it rains, any empty Brazil nut fruits on the forest floor fill with water and become home to several species. Scientists have identified a frog that breeds only in these water-filled husks. What caused these organisms to become so highly specialised, with their own continuing survival dependent on one species?

Figure 4.1 The Brazil nut tree, agouti rodent, euglossine bee and orchid are all found in the Amazonian rainforests of South America.

So many different species!

The Brazil nut tree, agouti, euglossine bee and orchid are all part of a tremendously rich **biodiversity** found in rainforests. We've got a multitude of different examples of organisms and species in this topic, reflecting the diversity of life on Earth. Today's biologists continue to collect, quantify and classify the living world. Yet despite their best efforts, we have only a vague idea of how many millions of species share the planet with us.

Recent estimates suggest there are over 30 million different species on Earth. Where does all the rich variety of living organisms come from, all so beautifully adapted to their surroundings? Charles Darwin and another naturalist, Alfred Russell Wallace, formulated a theory to answer this fundamental question, the theory of **evolution** by **natural selection**.

Biodiversity in crisis

There is growing concern about the number of species that are threatened with **extinction**. Why should we be concerned? Extinctions have occurred since life first appeared on Earth over 3500 million years ago. But the current extinction rate is at least 70 times faster than expected from the geological record, and is accelerating.

Forests and other ecosystems are valuable in their own right. And they are worth far more to us humans alive than when they are dead. Wild organisms provide all kinds of useful resources, vital to our survival and wellbeing. Organisms don't do these things for our benefit. For them it is just a matter of 'survival of the fittest' and we take advantage of their ingenuity.

Activity

Darwin discovered numerous new species, many of them in the Galapagos Islands, lying 1000 km off the west coast of South America. In **Activity 4.1** you can take a tour of the Galapagos Islands to see the variety of life that Darwin encountered. **A4.01S**

Activity

Activity 4.2 lets you develop some of the observation and interpretation skills that Darwin used and that today's biologists still use. **A4.02S**

Overview of the main biological principles covered in this topic

In this topic you will study how the huge biodiversity found on Earth came about. You will first look in detail at how species are adapted to different ecological niches. You then consider how they become adapted by means of natural selection. To understand what is meant by biodiversity, you will study how it can be classified and measured. You will also look at how classification has changed as new information is evaluated, and how DNA technology is helping to improve our understanding of the evolutionary relationships between living organisms.

You will consider the value of biodiversity in providing natural resources, focusing on the uses we make of plants. You will see that humans have relied on plants throughout history, and that the use of plant resources can contribute to sustainable living.

Building on what you have studied in earlier topics, you will compare the structure and function of starch and cellulose, and contrast plant and animal cell structure. You will examine specialisation of cells for support and transport through plant stems. You will look at how plant adaptations allow them to cope with physical hardships and the risk of being eaten.

Finally, you will examine the role of zoos and seed banks in the conservation of endangered species.

Review

Are you ready to tackle Topic 4 *Biodiversity and natural resources*?

Complete the GCSE review and GCSE review test before you start.

4.1 Why are there so many different species?

Every available habitat on Earth is colonised by living organisms, from frozen ice fields on high mountains to tropical rainforests. Somewhere between 1.4 and 1.7 million **species** have already been described and named (Figure 4.2).

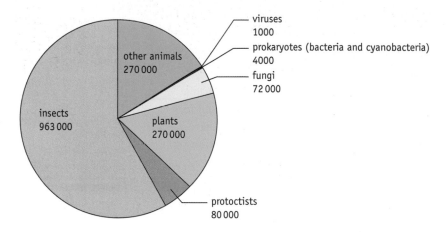

Figure 4.2 Estimates of the number of species currently known. It is thought that there are large numbers yet to be discovered – particularly insects, bacteria and fungi.

Key biological principle: What is a species?

What is a species? Perhaps surprisingly, this is one of the most difficult questions that biologists try to answer. As you can see from Figure 4.3, it is not enough to say that a species is a group of individuals that closely resemble one another. The most widely used definition is:

A species is a group of organisms with similar morphology, physiology and behaviour, which can interbreed to produce fertile offspring, and which are reproductively isolated (in place, time or behaviour) from other species.

Although horses and donkeys have, to some extent, similar morphologies (features), physiologies and behaviours, and can be bred together to produce mules, the mules are infertile. This means that, although horses and donkeys are closely related, they remain distinct species. However, some species do not seem to know this rule, and will quite happily breed and produce hybrid offspring which are fertile. The thirteen finch species of the Galapagos Islands have been known to interbreed and produce viable offspring. If this happened frequently we would expect small islands such as Daphne, with only four species of finch, to end up with birds of average types and no distinct species. But the finch species do

Figure 4.3 This is not a mouse! It is a marsupial and is more closely related to kangaroos and koalas than to rodents like mice or agoutis. Similar-looking organisms are not necessarily closely related.

remain distinct, because the hybrid forms do not survive drought conditions as well as the pure-bred types. There is enough fluctuation in the island's climate to maintain the different species.

Biologists sometimes disagree about a particular species, and sometimes one species is reclassified into two. Marsh and willow tits were once thought to be the same species (Figure 4.4). In 1999 the common pipistrelle bat was

CONTINUED ▶

split into two species on the basis of differences in their echolocation calls. The common pipistrelle (*Pipistrellus pipistrellus*) calls at a frequency of 45 kHz, while the soprano pipistrelle (*P. pygmaeus*) calls at 55 kHz.

Figure 4.4 The willow tit (left, *Parus montanus*) and marsh tit (right, *Parus palustris*) were thought to be just one species until about 1900.

Sometimes two separate species are reclassified as a single species. For example, the locust species *Locusta migratoria* and *Pachytylus migratoria* were merged into one species, known as the migratory locust (*Locusta migratoria*).

Modern techniques of DNA analysis are increasingly important in the identification of species. DNA analysis can help to tell us whether two slightly different populations are one species or two. For example, the British northern brown argus butterfly was thought to be a different species from Scandinavian populations, but analysis of mitochondrial DNA shows that they are the same species, *Aricia artaxerxes* (Figure 4.5).

DNA 'bar codes' are being developed to identify species without the need to look at their phenotypes. These work rather like the DNA fingerprints used in forensics, which you will study in Topic 6. In the future it may be possible to extract DNA from a plant in the field and identify it on the spot by comparison with a database using a satellite link.

However, DNA cannot always give a clear answer about species, because it cannot tell us whether two populations can breed together to produce fertile

Figure 4.5 A brown argus butterfly, *Aricia artaxerxes*.

offspring. We have to observe the two living populations in natural conditions in order to find out whether they can do this.

The decision about whether a population is one species or two is ultimately a matter of judgement. As Darwin found, isolated populations, such as those found on islands, often show differences from those on the mainland.

Q4.1 Read the descriptions below and decide if the groups are the same species or different species. Explain your answers.

a Group A has 98% of its genes in common with group B.

b Group C can interbreed with group A and produce fertile offspring.

c Group D is physically very similar to group B, except that it is a different colour. Where groups B and D overlap in geographical areas, mating between them produces offspring of an intermediate colour. The males of these offspring are infertile.

d Group E looks very similar to group F but has darker feathers. The genetic fingerprints of groups E and F are identical.

Species occupy different niches

Habitats and communities

Take a walk through a rainforest or a local wildlife area and you will soon see that the area is not uniform. There can be many different **habitats**. A habitat can be thought of literally as the place where an organism lives. A moorland, for example, is made up of undulating heather-clad slopes, streams, bogs, and rocky outcrops. Each habitat has a particular set of conditions which supports a distinctive combination of organisms.

Within a habitat there may be many populations of organisms. Each **population** is a group of interbreeding individuals of the same species found in an area. The various populations in an area make up a **community**. So, for example, a rainforest community consists of all the organisms that live in the rainforest, which might include the Brazil nut trees, agoutis, euglossine bees, frogs and orchids.

An organism's environment provides all its essential resources such as energy, raw materials, a place to live and a mate. Most resources needed for survival are in limited supply. In order to survive and reproduce, organisms must compete successfully for the resources they need, against other organisms of the same and different species.

Different niches avoid competition

Two species sharing the same habitat tend not to be in competition with each other. Each species occupies a particular **niche**, often referred to more fully as an ecological niche. A niche can be defined as 'the way an organism exploits (uses) its environment'. All the species sharing a habitat have different niches. If two species live in the same habitat and have exactly the same role within the habitat – the same food source, the same time of feeding, the same shelter site, and so on – they occupy the same niche and they will compete with each other. The better-adapted organisms will out-compete the other and exclude it from the habitat. This happened to the red squirrel when the North American grey squirrel was introduced to England in the nineteenth century. The native red squirrel has been out-competed and replaced by the grey squirrel in much of England.

Orchid niches

The niches of some orchids exploit the behaviour of insects as part of their highly specialised pollinating mechanisms. As you saw in the introduction to this topic, female euglossine bees visit the flowers of Brazil nut trees to collect nectar. At the same time they inadvertently pick up and carry pollen from one tree to the next, pollinating the flowers. The male euglossine bees visit orchids to feed and to collect aroma chemicals, which are important in attracting a female mate.

Pyramidal orchids, a common sight on chalk and limestone grassland across England and Wales, produce scent which attracts day-flying moths. As a moth probes into the deep, tapering corolla of the orchid flower, a parcel of pollen in the shape of a collar attaches itself with quick-drying glue around the moth's tongue. After a short time, the sides of the pollen mass curl forward. The pollen is now in just the right position to pollinate the two stigmas of the next flower the moth visits.

Some orchids are so specialised that they can only be pollinated by one species of insect. For example, the mirror orchid (*Ophrys speculum*) is pollinated by just one species of wasp. Only male wasps visit the orchid, which resembles a female wasp of that species. The male mistakes the flower for a female wasp and, as he attempts to mate, the pollen mass sticks to his head. He then carries this to the next flower as he attempts to mate again. The flower even produces a scent similar to the pheromone made by the female wasp to attract males. About

Figure 4.6 Digger wasp 'mating' with a fly orchid flower. When the female wasps emerge two weeks after the males, the flowers are no longer visited.

400 species of orchid worldwide are known to mimic female insects and trick their pollinators into 'pseudocopulation' (Figure 4.6). By doing so they out-compete other flowering plants for the attentions of insect pollinators.

Q4.2   Suggest **a** an advantage, and **b** a disadvantage of using just one type of insect pollinator.

Woodpecker niches

There are three species of woodpecker in Britain, the lesser spotted, the great spotted and the green woodpecker. All woodpeckers have a powerful beak which they use for probing into rotting wood. They pick up food such as woodlice, beetle larvae and ants using a very long tongue. The tiny lesser spotted woodpecker prefers the finer branches at the tops of trees, whereas the great spotted tends to feed on the broader branches (Figure 4.7). The green woodpecker tends to feed on the ground. All three species can co-exist in the same habitat because each has a different niche.

Woodpeckers occur throughout most of the world, but they are not good at flying over large stretches of ocean, and have never reached Australasia, Madagascar or the Galapagos Islands. In Madagascar the woodpecker niche is occupied by the aye-aye, which uses its long bony middle finger in a similar manner to a woodpecker's beak. In the Galapagos Islands, off the coast of Ecuador, no species has yet evolved a comparable structure to the woodpecker's beak. However, one small bird, called the woodpecker finch, has acquired the habit of breaking off a cactus spine and using it to probe into rotting wood, just like a woodpecker.

Q4.3   What do you think might happen to the woodpecker finch if a pair of breeding woodpeckers arrived in the Galapagos Islands?

Activity

In **Activity 4.3** you can work out how one bee is exploiting its niche.
A4.03S

Figure 4.7 Different feeding strategies ensure that this great spotted woodpecker does not compete with other woodpecker species.

Adapted to their environments

Woodpeckers, aye-ayes, orchids and all other organisms are **adapted** in such a way that they are uniquely able to exploit their own particular niches. Being adapted means being specialised to suit the environment in which the organism lives. Features which enable organisms to survive are called **adaptations**. Adaptations can be classified as behavioural, physiological or anatomical, though these are not hard and fast categories – there is some overlap.

Behavioural adaptations

Behavioural adaptations are any actions by organisms which help them to survive or reproduce. When agoutis bury Brazil nuts, this is a behaviour which will aid their survival by providing a future food source. Even plants respond to their environments, for example as they turn their leaves towards the Sun, maximising the amount of light received for photosynthesis. The explosive dispersal of plant seeds, for example in balsam and bitter cress, could also be described as a behavioural adaptation. Some fungi use explosive mechanisms to expel their spores. The Costa Rica palm, *Socratea durissima*, has stilt-like roots which grow towards a sunny patch and slowly pull the whole tree towards it.

Q4.4 For each of the behaviours described below, suggest what the organism is responding to and how the behaviour helps it survive.

a Herring gull chicks peck at the red spot on the parent's beak, causing the parent to regurgitate food.

b Some ground-nesting birds, such as ringed plovers, perform a 'broken wing display' on seeing a predator (Figure 4.8). The parent bird runs away from the nest, trailing its wing as if it is broken. But just as the predator is approaching, the bird takes off and escapes.

c In drier conditions, woodlice move around more. In humid conditions they move about much less.

Figure 4.8 Broken wing display by ringed plover.

Physiological adaptations

Physiological adaptations are features of the internal workings of organisms which help them to survive or reproduce.

British motorists in spring-time may notice a swathe of tiny white flowers along the very edges of the kerb. The flowers are Danish scurvy grass, *Cochlearia danica*, a member of the cabbage family. It was formerly found mainly by the sea, but has spread inland with the increasing use of salt on icy roads (Figure 4.9). The plant has a physiological adaptation allowing it to tolerate high salt concentrations. It has therefore been able to occupy a newly created niche unavailable to other inland plants.

Activity

Investigate seed beetle behavioural adaptations in **Activity 4.4. A4.04S**

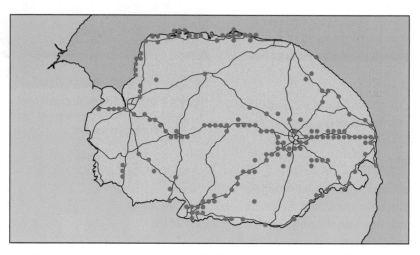

Figure 4.9 The distribution of Danish scurvy grass, *Cochlearia danica*, in Norfolk. Red lines show main roads.

Q4.5 Humans are unable to survive by drinking sea water, but whales, fish and many marine birds are able to take in sea water and excrete the excess salt. Salt excretion is achieved by ion pumps in kidneys, fish gills and in the beaks of some birds. Name the process by which salts are pumped across membranes using ATP energy.

Hot springs make a very extreme environment, yet some organisms are adapted to them physiologically. Thermophilous bacteria can tolerate the highest temperatures of any organism. Those found in thermal vents deep in the ocean can survive at 350 °C. Others live in hot springs, such as those at Yellowstone National Park, at temperatures up to 80 °C. They achieve their heat tolerance by having heat-stable enzymes. Such enzymes have useful applications for humans, such as the production of biological washing powders.

Anatomical adaptations

Anatomical adaptations are the structures we can see when we observe or dissect an organism. For example, the bodies of bumblebees show adaptations used to collect nectar and pollen. There are about 250 species of bumblebee worldwide. Each has a long tongue through which it can suck nectar (sugary solution) from flowers. Within a habitat, some bumblebee species have longer tongues than others, so they are specialised to feed from different flowers.

Q4.6 Table 4.1 shows the flowers visited by six species of bumblebee seeking nectar in an English garden. The flowers are listed in order of increasing corolla length (length of petal tube), and the bumblebees in order of increasing tongue length. Explain what the data show about the niches of these six bumblebee species, when competing for nectar in this garden.

	Corolla length/mm	*Bombus pratorum*	*Bombus lapidarius*	*Bombus lucorum*	*Bombus terrestris*	*Bombus pascuorum*	*Bombus hortorum*
Tongue length/mm		5	6	7	8	13	15
Mallow	0						
Daisy	0						
Poppy	0						
Yellow clover	1						
Shepherd's purse	2						
Bramble	3						
White clover	3						
Bird's foot trefoil	4						
Snowberry	4						
Gooseberry	4						
Broccoli	5						
Woundwort	8						
Delphinium	8						
Foxglove	10						
White dead-nettle	11						
Sage	12						
Sweet William	15						
Broad bean	15						
Honeysuckle	35						

Table 4.1 Flowers visited by six species of bumblebee (*Bombus*) with increasing tongue length.

Activity

In **Activity 4.5** discover more animal and plant adaptations, and listen to Sarah Darwin talking about the Galapagos tomato. Endemic to the islands, it is a salt-resistant plant able to grow in salty soil. **A4.05S**

Bumblebees also have a pollen basket on their hind legs. This is a fringe of hairs which holds pollen combed from the body (Figure 4.10), allowing pollen to be carried back to the nest where it will be used to provide food for the larvae. A slot on the front legs is used by a bumblebee to clean any pollen grains stuck on its antennae.

Q4.7 Describe one adaptation in each of the species in the Brazil nut 'story' at the start of the topic, and in each case state if it is a behavioural, physiological or anatomical adaptation.

As you saw in the Brazil nut and orchid bee example, plants and insects can evolve in tandem. The plant and its pollinator become dependent on each other and more and more closely adapted. This is called **co-adaptation**.

Figure 4.10 Bumblebee (Queen *Bombus lucorum*) with full pollen baskets.

Another example of co-adaptation is the whistling thorn, found on African savannahs. The plant grows swellings with entrance holes on its stems. Ants make their home inside the swellings, and swarm out to deter browsing animals such as giraffes when they try to eat the plant's leaves.

Q4.8 Suggest how the ants benefit from living in the swollen stems of the whistling thorn described above.

✔ Checkpoint

4.1 The structures of roots, leaves, stems, flowers and fruits all provide beautiful examples of anatomical adaptations in plants. Describe one way that each is adapted and suggest how the adaptation aids the plants' survival.

4.2 How did organisms become so well adapted?

Natural selection

Brazil nut trees, agoutis, orchids and all other species are adapted in such a way that each is uniquely able to exploit its own ecological niche. **Natural selection** is the mechanism, first proposed by Darwin and Wallace, by which organisms change over time as they adapt to their changing environments. As a population increases in size, a greater proportion of individuals will die or fail to reproduce owing to competition for resources, such as food and space. Disease and extreme environmental conditions may also cause the deaths of some individuals. This striving for survival is known as the 'struggle for existence'.

Survival of the fittest

In this struggle for existence there will be winners and losers. Winners are those individuals which, by chance, possess some characteristic which gives them an advantage over others. This differential survival is called 'the survival of the fittest'. (Here, 'fit' means well adapted, rather than physically fit in an athletic sense.) In the case of orchid bees it might be the possession of a slightly longer tongue, which improves their ability to compete for food. They have a selective advantage over other orchid bees, and are more likely to survive and reproduce, passing on the alleles for a longer tongue to their offspring. The average tongue length of the next generation of bees will be slightly greater. Such a change in form (or behaviour or physiology) over generations is called **evolution**.

Evolution by natural selection

Evolution is more precisely defined as 'a change in allele frequency in a population over time (generations)'. In our orchid bee example, there has been a change (increase) in the relative frequency of alleles for a long tongue. The emphasis on alleles in this definition reminds us that, for natural selection to lead to evolution, there must be some genetic variation in the population. An allele can be selectively neutral (i.e. has no advantage or disadvantage) but suddenly become selectively very advantageous when the environment changes.

Q4.9 Would you expect bee tongues to go on getting longer and longer with each generation as a result of natural selection? Give a reason for your answer.

The fundamentals of the theory are summarised in Table 4.2.

1	A population has some naturally-occurring genetic variation with new alleles created through mutations.
2	A change in the environment causes a change in the selection pressures acting on the population.
3	An allele which was previously of no particular advantage now becomes favourable.
4	Organisms with the allele are more likely to survive, reproduce and so produce offspring.
5	Their offspring are more likely to have the allele, so it becomes more common in the population.

Table 4.2 Evolution by natural selection.

Activity

Activity 4.6 lets you investigate natural selection. A4.06S

Extension

You can find out how to calculate allele frequencies in Extension 4.1. X4.01S

Natural selection is happening right now!

Natural selection is occurring all the time, but it can be very slow and difficult to observe during a human lifetime. If the environment remains stable, then organisms may simply become better and better adapted to their existing niches, as Figure 4.11 shows. But if the environment changes, then selection changes too. For example, our use of pesticides and antibiotics has led to the evolution of resistant forms of insects and bacteria. One example you may be familiar with is evolution within head lice (Figure 4.12).

In this example the lightest and darkest phenotypes are more conspicuous to prdators

Figure 4.11 Stabilising selection.

About 7% of UK primary school children have an infestation of head lice in their hair. These parasites suck blood and have a rapid rate of reproduction, doubling in number every seven days. There are many head lice shampoos which contain insecticides. Permethrin and malathion are frequently the active ingredients. However, in recent years there has been both anecdotal and scientific evidence to show that head lice have become resistant to these shampoos and that the shampoos are no longer as effective. Results of a study carried out at the University of Bristol showed that there was an 87% failure rate for permethrin and a 64% failure rate for malathion in the treatment of infested school children.

This is a modern example of natural selection in action. Initially, many head lice must have been killed by the insecticides, but now it seems that permethrin will kill only 13% of them. Most head lice survive and quickly breed the next resistant generation.

In this example, the insecticides in the shampoos provide the selection pressure. Had there not been pre-existing variation in the population, all of the lice on treated heads would have died. Without those original chance mutations the head louse could have become extinct – at least in areas where the chemical treatments against it were used.

Figure 4.12 Coloured scanning electron micrograph (SEM) of a human head louse, *Pediculus humanus-capitis*. An egg (nit) has been laid on the hair (lower right). The lids of the nits (top) are perforated with air holes that allow the developing nymphs to breathe. Adult female lice lay between 80 and 100 eggs in a lifetime.

Evolution has occurred: the insecticide-resistance alleles in head lice went from being quite rare in pre-insecticide populations to being extremely common.

In the case of the resistant head lice, evolution by natural selection has occurred, producing a population which has different allele frequencies from the original population.

Q4.10 The pictures in Figure 4.13 below show the events in a hypothetical instance of evolution. Match the statements in Table 4.2 with the correct pictures.

Checkpoint

4.2 Identify three key steps in the process which resulted in resistant head lice.

Figure 4.13 Evolution in action.

Being adaptable

The ability of a population to adapt to new conditions will depend on:

- The strength of the selection pressure
- The size of the gene pool
- The reproductive rate of the organism.

Insects, such as mosquitoes, have rapidly become resistant to insecticides such as DDT, because (i) the selection pressure is strong (DDT originally killed more than 99% of all mosquitoes) (ii) there are millions of mosquitoes with a very large gene pool and (iii) mosquitoes reproduce very quickly, so any surviving (resistant) ones can quickly build up numbers, passing on their resistance to their offspring.

Q4.11   What might cause a small number of mosquitoes to be naturally resistant to an insecticide?

The current rate of climate change due to global warming is an example of selection pressure. Some organisms will survive and some will not. Topic 5 in the A2 course discusses the implications of global warming for species that are and are not adaptable to it.

Gene pools

A **gene pool** consists of all the alleles of all the genes present in a population.

Imagine all the people in your school or college taking out a set of their genes and throwing them into one big box. The box would then contain the gene pool of the school/college population. In the gene pool box there will be the different alleles of each gene (of the 20 000 or so) thrown in. Some alleles of each gene will be common in the gene pool and others will be rare.

The concept of a gene pool is useful when thinking about the biodiversity and adaptability of any species. Populations with a bigger gene pool (more different alleles of each gene) are said to have greater genetic diversity. New alleles are produced all the time by mutation of existing alleles, but this is a slow and random process. When the population of a species declines too far, some alleles are lost, and the genetic diversity (biodiversity) of the species declines.

Q4.12   The use of insecticides in agriculture in the 1960s led to the poisoning of some birds of prey, such as peregrine falcons and sparrowhawks. Suggest why these birds did not develop resistance to the insecticides.

Q4.13   Explain why it is important for the survival of large mammals, such as rhinos and whales, to conserve a large and varied gene pool.

Being perfect

Can living organisms become perfectly adapted to their environments? The answer is no, for the following reasons. When the environment changes, there is a time lag before a population can adapt by natural selection, so organisms are always playing a game of evolutionary catch-up. Even in a stable environment, mutations are occurring all the time, many of which are harmful to the organism. This makes a population less well adapted to its surroundings.

Being perfect, or close to perfect, has its benefits and drawbacks. Becoming adapted to a specialised niche helps avoid competition with other species living in the same area. The specialisation of the species involved in the Brazil nut 'story' means they avoid competition for food, breeding ground or other features of their niche. However, their 'perfection' also makes them interdependent and vulnerable; each species would be severely affected if any one of the others became rare. On the other hand, species which can adapt to a wide range of habitats will always be in competition with a range of other species, but they are less vulnerable to changes in their habitat.

Did you know?

Nature thought of it first

Many human inventions – including antifreeze, flight, warning colours, sensors and Velcro – have their counterparts in nature. We sometimes say that 'nature thought of it first'. The idea for Velcro came from noticing how hooks on fruits enable them to stick in animals' fur, assisting the dispersal of seeds (Figure 4.14).

Figure 4.14 Scanning electron micrograph of a hooked barb on the fruit of cleavers or goosegrass, *Galium aparine*, grasping a strand of Shetland wool.

Of course it is not quite right to say that 'nature' thought of something. It would suggest, for example, that mosquitoes invented resistance in order to survive pesticides. This kind of thinking, which implies a purpose behind evolution, is called teleology. Teleology is bad biology, but it is difficult not to slip into teleology when explaining the complex adaptations of living organisms. Remember that adaptations come about by natural selection of random genetic variation, and not by forward planning.

4.3 Quantifying biodiversity

What is biodiversity?

'Biodiversity' is a shortened form of the words 'biological diversity'. Biodiversity is used in everyday language to mean the variety of life, and in particular the wealth of different species that exists as a result of evolution by natural selection. In biological terms it refers to the variety of species that belong to every different group of organisms, animals, plants, fungi, bacteria and other microorganisms, living in all the habitats on the planet. Biodiversity also refers to the diversity within species.

Current estimates of the actual number of species on Earth range from 5 million to 30 million or more. In just one animal group alone, the beetles, there are thought to be at least 1 million and perhaps as many as 5 million species. So far, only about 450 000 beetle species have been described and named, less than half the world's beetle diversity. The situation is probably far worse in other less well studied or less popular groups such as bacteria and fungi, where only a tiny fraction of the species have been given a name. We still have no rigorous estimate of how many species we share the planet with. This is not because of any particular technical difficulties. It's simply because the size of the task is massive.

Biologists want to be able to quantify biodiversity. Once biodiversity has been assessed, the data can be mapped on both a local and a global scale to reveal patterns in diversity. This can help to focus conservation efforts on vulnerable habitats or species. But to do this and to understand the Earth's biodiversity, biologists need to define what is meant by biodiversity and be able to measure it. This is only possible if they can identify and classify the species observed.

Activity

In **Activity 4.7** you can consider different ways of defining biodiversity. **A4.07S**

Activity 4.8 lets you read about biologists involved in biodiversity research and helps you investigate the rate at which new species are discovered. **A4.08S**

Extension

In **Extension 4.2** read about where estimates of species numbers come from. **X4.02S**

? Did you know?

Producing a species catalogue

The international Species 2000 Programme was set up in September 1994 with the objective of making a catalogue of all the known species of plants, animals, fungi and microbes on Earth as the baseline dataset for studies of global biodiversity. By 2000 its 'catalogue of life' contained 220 000 species. This had risen to one million by 2007, which is a sizeable proportion of the estimated 1.75 million species that have been described and named. The programme was established by the International Union of Biological Sciences (IUBS) and is based in the School of Plant Sciences, University of Reading.

The efforts of the Species 2000 Programme may seem impressive compared with the mere 12 000 or so species described by Carolus Linnaeus. Linnaeus, a great Swedish biologist, made the first attempt to produce such a global species catalogue (*Systema Naturae*, published in 1758). However, the task still facing the Species 2000 Programme remains huge. This means that a great deal more exploration, discovery and naming has still to be done by biologists.

New technology will help speed up the process. Geographical positioning systems (GPSs) can quickly pin-point the location of new species whilst the Internet can be used to store and exchange data from around the world. Crucially, DNA analysis can now help to confirm that 'new' species are indeed different from existing species.

Unlocking identities

Putting a name to the species they find represents a significant challenge to biologists working in the field. Whether they are a research worker studying agoutis in a tropical rainforest, or a bird watcher looking at wildfowl on an estuary in the UK, accurate biological identification is vital.

Unique names

All organisms are given a scientific name. This avoids the confusion that can arise when common names are used. For example, *Arum maculatum* (Figure 4.15) has more common names than any other plant in Britain, including lords and ladies, and cuckoo pint. The Brazil nut is known locally as the castana; its scientific name is *Bertholletia excelsa*. Probably the most important thing that Carolus Linnaeus, the Swedish biologist, did in his early attempts to catalogue living things was to come up with a system in which each species was given a unique two-part Latin name. This **binomial system** is still in use today. The first part of the name, the **genus**, is shared by all closely related species, so all horses and zebras are in the genus called *Equus*. The second part of the name defines the particular species in the genus. Together these two words make up a unique species name that is often highly descriptive (Figure 4.16).

Figure 4.15 More than a dozen common names have been recorded in Britain for *Arum maculatum*. Species names are always written in italics or underlined. The first part of the name, the genus, always starts with a capital letter. The second part of the name does not.

Figure 4.16 This bat has the scientific name *Rhinolophus ferrumequinum*, from *rhinos* (Greek for nose), *lophos* (Greek for crest), *ferrum* (Latin for iron) and *equinum* (Latin for horse), providing an excellent description of its nose – if you understand Greek and Latin. Its common name is the greater horseshoe-nosed bat.

Q4.14 Using the common and scientific names below, **a** State which genus creeping buttercup belongs to; **b** Decide whether meadow sweet and dropwort, or lady's mantle and dropwort, share more similarities, and give a reason for your answer.

Common name	Scientific name
creeping buttercup	*Ranunculus repens*
meadow sweet	*Filipendula ulmaria*
lady's mantle	*Alchemilla vulgaris*
dropwort	*Filipendula vulgaris*

Did you know?

Extension

Use the SAPS computer-based key in Extension **4.3. X4.03S**

Biological identification

To name an organism you have to be able to identify what it is. Keys are a traditional means of identifying organisms. You will no doubt be familiar with the use of a simple text-based **dichotomous key**, which means that there are always two alternatives at each stage in the key. Versions of this type of key are routinely used by field biologists to find the names of unfamiliar organisms.

Traditional dichotomous keys have to be used in a set sequence, starting with the first pair of statements. Sometimes this can be a problem because, if a characteristic referred to early in the key is absent due to damage or the stage of development of the organism, it is difficult to continue to use the key. Multiple access keys are more flexible. They allow the user to start with whichever characteristic they wish, and work through the key in a different order.

Often regarded as one of the most traditional areas of biology, identification methods are increasingly making use of new technology. The field of computer-assisted taxonomy (CAT), which uses computer-based keys and electronic techniques, is expanding and beginning to replace paper-based keys. Computer-based keys can allow multiple access and easily be updated when new species are discovered.

Q4.15 Use the key in Figure 4.17 to identify the butterfly fish shown in Figure 4.18.

Decision point	Choices	Decision
1	Pelvic fin dark	2
	Pelvic fin light	4
2	Two large white spots below dorsal fin	*C. quadrimaculatus*
	Lacks two large white spots below dorsal fin	3
3	Tail with two dark bars at tip	*C. reticulatus*
	Tail with one dark bar at tip	*C. kleinii*
4	Posterior or dorsal fin has long filament extension	5
	Filament extension lacking from dorsal fin	6
5	Large dark spot on body near filament	*C. ephippium*
	Small dark spot on body near filament	*C. auriga*
6	No vertical band through eye	*C. fremblii*
	Vertical band through eye	7
7	Incomplete eyeband on face (does not go to top of head)	*C. multicinctus*
	Complete eyeband on face (extends to top of head)	8
8	Nose area with band	9
	Nose area lacks band	10
9	Fewer than eight diagonal bands on body	*C. ornatissimus*
	More than eight diagonal bands on body	*C. trifasciatus*
10	Distinct white spots splits eyeband above eye	*C. lineolatus*
	No white spot above eye; eyeband not split	11
11	Upper third of body under dorsal fin dark	*C. tinkeri*
	Upper third of body under dorsal fin not dark	12
12	Distinct small spots arranged in rows	13
	No distinct small spots; body has large spot or band	14
13	No black band on caudal fin	*C. citrinellus*
	Obvious black band on caudal peduncle	*C. miliaris*
14	Side with a large black teardrop; no dark bars on tail	*C. unimaculatus*
	Large black shoulder patch; tail with dark bars	*C. lunula*

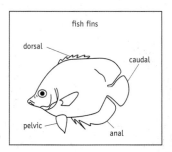

Figure 4.17 Dichotomous key for butterfly fish of the genus *Chaetodon*.

Figure 4.18 Use the key in Figure 4.17 to identify this butterfly fish.

Sorting and grouping

A hierarchical system

Faced with such a bewildering variety of living things, humans have always tried to organise and make sense of the variety of life. Placing organisms into groups based on shared features, known as classification or **taxonomy**, results in a manageable number of categories and has been the principal aim of all classification systems. Linnaeus created the first classification system, grouping organisms according to their visible similarities and differences.

On coral reefs there is often an amazing array of colourful fish. But on closer inspection, similarities between groups of fish become apparent. Look at the butterfly fish in Figure 4.19. What features do they all have in common? They are thin-bodied fish, many having a dark band across the eye. Each has a small mouth at the end of an extended snout, and tiny bristle-like teeth. The three butterfly fish in Figure 4.19 all have these features and all belong to the genus *Chaetodon*. Look at Figure 4.19 and decide if you can see the similarities.

Figure 4.19 A The archer butterfly fish (*Chaetodon bennetti*), **B** the white spotted butterfly fish (*Chaetodon kleinii*) and **C** the limespot butterfly fish (*Chaetodon unimaculatus*). These butterfly fish share common characteristics and all belong to the same genus.

Q4.16 Look at Figure 4.20. What distinctive feature do the bannerfish in the genus *Heniochus* seem to have in common?

Figure 4.20 Bannerfish are related to the butterfly fish in Figure 4.19, but are sufficiently different to be classified in a different genus.

The bannerfish shown in Figure 4.20 look rather like the butterfly fish in Figure 4.19, they are sufficiently different to be grouped in a different genus, *Heniochus*. However, they do share a number of similarities with members of the genus *Chaetodon*, like the butterfly fish, so the two genera are grouped together in the same **family** – the Chaetodontidae (butterfly fishes).

Together with many other reef fish, members of this family all have thoracic as opposed to abdominal pelvic fins (that is, their lower fins are towards the front of the body). All families sharing this feature are grouped in the **order** Perciformes (perch-like), the largest order of fish, and all the orders of fish with bony skeletons are grouped together in the **class** Osteichthyes (bony fish). This class, along with all the other classes of animals with a dorsal spinal cord, is placed in the **phylum** Chordata – see Figure 4.21. All animals are grouped in the **kingdom Animalia**.

A taxonomic hierarchy is created. This is a series of nested groups or **taxa** (singular **taxon**), in which the members all share one or more common features or **homologies**. It is similar to the way in which you might arrange the folders and files in your work area of a computer (assuming you are that organised!).

The hierarchy of groups is:

- kingdom
- phylum (plural phyla)
- class
- order
- family
- genus (plural genera)
- species.

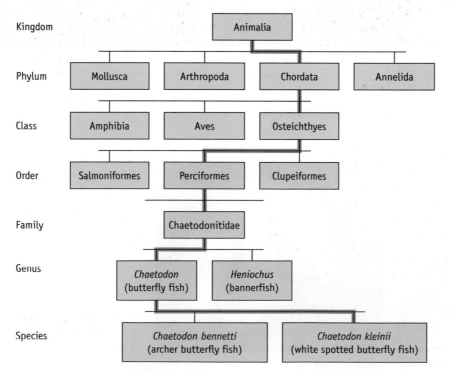

Figure 4.21 Butterfly fish classification from kingdom to species (don't worry – you don't need to remember the names!)

Butterfly fish are clearly animals. Along with many other organisms, we readily recognise them as members of the kingdom Animalia. In the same way, we can easily recognise the sea grass growing in the shallow waters around a coral reef as a member of the kingdom **Plantae**. Initially taxonomists recognised only these two kingdoms; Plants and Animals. As scientists acknowledged that all organisms did not fit into these two kingdoms, further groups were formed. The separation of living things into five main kingdoms was proposed by R. H. Whittaker in 1959 and later modified by L. Margulis and K. V. Schwartz in 1982, giving the five kingdoms **Animalia**, **Plantae**, **Fungi**, **Protoctista** and **Prokaryotae**.

- Animalia kingdom – multicellular eukaryotes that are heterotrophs, organisms that obtain energy as 'ready-made' organic molecules by ingesting material from other organisms.

- Plantae kingdom – multicellular eukaryotes that are autotrophs, organisms that make their own organic molecules by photosynthesis (except a few parasites).

- Fungi kingdom – multicellular eukaryotes that are heterotrophs, which absorb nutrients from decaying matter after external digestion.

- Protoctista kingdom – eukaryotes that photosynthesise or feed on organic matter from other sources but are not included in the other kingdoms, includes single-celled protozoa, such as *Amoeba* and *Paramecium*, and algae.

- Prokaryote kingdom – prokaryotic organisms, includes the bacteria and blue-green bacteria (Cyanobacteria).

As well as being a logical way to organise biodiversity, such as the shoals of brightly coloured fish around a coral reef, the groups have a deeper biological significance. Since members of a taxon share common features, it is likely that they also share a common evolutionary ancestor. This means that, in evolutionary terms, members of a taxon will be more closely related to each other than they will be to any organisms outside that taxon. In terms of the butterfly fish, members of the genus *Chaetodon* are all more closely related to each other than to members of the genus *Heniochus*, and all of these fish are more closely related to one another than they are to members of other fish families. Diagrams of classification tell an evolutionary story, rather like a family tree.

A new system developed

In the 1960s, a scientist called Carl Woese, working at the University of Illinois, aimed to define the evolutionary relationships of prokaryotes. Woese pioneered RNA sequencing of bacteria, using the techniques available at the time. He used the sequences to construct phylogenies of groups of bacteria (Figure 4.22). A decade later, he noticed that one complete group of bacteria, the methanogens, completely lacked the sequences characteristic of bacteria. The RNA sequences showed up as different from all other bacteria, even with the crude early genetic sequencing techniques used by Woese. He supported his ideas with additional evidence, for example, the methanogens, unlike other bacteria, had no peptidoglycans (polymers of amino sugars cross-linked by peptide bonds) in their walls. They also had membrane lipids that differed from those of eukaryotes and other bacteria.

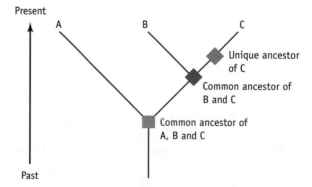

Figure 4.22 Phylogenetic trees show the evolutionary relationships between organisms based on molecular differences, the result of mutations accumulated over time.

A new group identified

In 1977, Woese proposed that this group belongs to a new category of organisms, the Archaea, forming a third branch of life alongside the prokaryotes and eukaryotes. Archaea may be related to some of the earliest life forms, existing before there was oxygen in the atmosphere. Today they survive in extreme anaerobic environments, in places such as hot springs and salty lakes.

A paper published in the *Proceedings of the National Academy of Sciences* announced the new group. Individual scientists or research groups report their findings by publishing their work in scientific journals, in conference papers and on the Internet. This allows other scientists to see and comment on the data collected, the methods used, and the interpretations. The scientific community carefully checks their data and interpretations of these data. This has to happen before new ideas are accepted as reliable scientific knowledge. This process of 'peer review' helps to detect invalid claims and adds weight to valid ones. But it can be difficult to get new theories accepted, particularly when they conflict with accepted theories.

Scientific community sceptical

Woese's theory was suggesting that most life is microbial and that scientists had been ignoring the existence of one of three major forms of life on Earth. Most of the scientific community were sceptical and ignored or dismissed Woese's ideas. Few said anything to Woese directly, or even responded in journals. Woese hated large meetings, so had few opportunities to argue in person on behalf of the Archaea. He seldom even attended the annual meetings of the American Society for Microbiology (ASM).

Evidence supports three domains

Luckily, the influential German microbiologist Otto Kandler accepted Woese's work. It was Kandler who organised the world's first Archaea conference in 1981, and influenced other scientists to accept the new group. He and other scientists published papers on the subject in respected international journals, and eventually the theory gained acceptance as more and more scientists reported evidence showing the Archaea to be unique. In 1990, Woese, Otto Kandler and Mark Wheelis proposed a new system of classification, a universal phylogenetic tree based on three **domains**, the Archaea, the Bacteria and the Eucarya (Figure 4.23). The organisms in each of the three domains contain RNA sequences that are unique to their domain.

The acceptance of the validity of Woese's classification was a slow process, but increasingly, as new evidence emerged as a result of modern techniques, the scientific community critically evaluated the evidence, accepted his theory and recognised its full significance.

The final ratification of Woese's work came with completion of the genetic sequencing of *Methanococcus jannaschii* in 1996, which confirmed that the Archaea do indeed form a unique domain, with a closer evolutionary relationship to eukaryotes than to bacteria.

Classification is dynamic

Phylogenetic trees represent the evolutionary relationships based on the best existing evidence and they are constantly being revised as new data becomes available.

Checkpoint

4.3 Produce a series of bullet points to summarise how new data was used to produce the three domain classification, highlighting the role of critical evaluation by the scientific community.

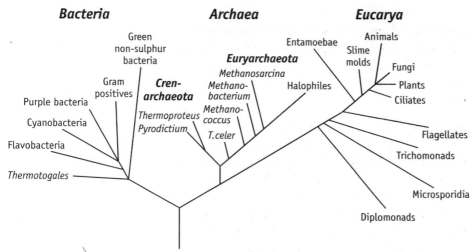

Bacteria

Archaea

Eucarya

Figure 4.23 In 1990, Woese, Otto Kandler and Mark Wheelis proposed a new system of classification based on three domains, the Archaea, the Bacteria and the Eucarya. It was derived from molecular differences caused by accumulation of mutations over time. By comparing the similarity of RNA between two species, it is possible to show their evolutionary relationship. Phylogenetic trees such as this suggest where species branched off from common ancestors.

All modern classification systems attempt to show evolutionary relationships between species. DNA analysis often supports traditional Linnaean classifications, but not always. For example, in the traditional classification of the vertebrates, birds and reptiles are placed in separate classes. Anyone can see that birds differ from reptiles. They are warm-blooded and possess feathers, beaks and hard-shelled eggs. But DNA analysis places birds in the same taxon as turtles, lizards, snakes and crocodiles (Figure 4.24). This new classification is supported by fossil evidence, which shows that birds are descended from a type of dinosaur.

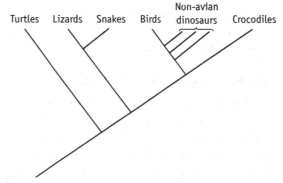

Figure 4.24 The group Reptilia, based on DNA analysis, includes animals from two separate Linnaean classes, reptiles and birds, showing that birds are as closely related to the different reptile groups as the reptiles are to each other.

Activity

In **Activity 4.10** you can investigate changing classification for yourself. **A4.10S**

Phenotypic traits are observable characteristics such as flower structure or bone structure. Organisms sometimes look similar only because they are adapted to similar conditions. For example, many plants living on windy mountain tops have hairy leaves (Figure 4.25), but they may belong to several different plant families. Dolphins and sharks look superficially similar, but they are not closely related (Figure 4.26). See also the mouse-like marsupial in Figure 4.3. These are examples of convergent evolution – unrelated organisms evolving a similar appearance as they adapt to the same habitat. In the case of the dolphin and the shark it is easy to see that they belong to different groups: a dolphin is a mammal and breathes air using lungs, whereas a shark is a

Figure 4.25 Scanning electron micrograph of leaf hairs of hairy mullen, typical of plants exposed to strong light and drying winds.

fish and has gills. But it is not always possible to be sure whether similarities are due to true relatedness rather than convergence. In addition, some organisms can have markedly different phenotypic traits from their close relatives. Using DNA analysis in classification overcomes these problems.

Figure 4.26 A dolphin and a shark share a similar shape, adapted to fast swimming and catching fish, but belong to different classes of vertebrates (mammals and fish).

Why classify?

Classification is a way of thinking about biodiversity. When used in partnership with genetic analysis, classification enables us to look at evolutionary relationships.

The ability to identify closely related species can be very valuable, for example when trying to find new sources of chemicals with medicinal or other beneficial properties. An anti-viral drug was identified in the Australian tree species, the Moreton Bay Chestnut, but was found to be toxic. However, plants in a closely related genus were found to contain more of the drug in a less toxic form.

Biodiversity within a species

Individuals within a species differ from one another – they show variation (Figure 4.27). You only need look around at the other members of your advanced biology group to realise just how much variation there can be. Topic 3 discussed how the appearance of an organism, its phenotype, is the result of interactions between its genotype and the environment. In all organisms that reproduce sexually, every individual (except for cases such as identical twins and cloned organisms) has a unique combination of alleles. This is **genetic diversity**, and the greater the variety of genotypes the more genetically diverse the population. Genetic diversity allows the population to adapt to changing conditions so should be conserved.

Figure 4.27 At first glance the zebras may all look the same, but on closer inspection there are differences between the individuals. Some of this variation will be genetic in origin.

Sources of genetic variation

Where does all this genetic diversity come from? Topic 3 explained how meiosis results in genetic variation through independent assortment and crossing over. Random mutations also generate genetic variation (Figure 4.28). They do this by changing the DNA sequence in the cells of an organism, creating new alleles. Genetic diversity is increased with the addition of these new alleles to the gene pool. The gene pool consists of all the alleles of all the genes present in a population. In Topic 2 you saw how the deletion of three nucleotides was one of the mutations causing cystic fibrosis. You also saw how a single point mutation (alteration of one base) could result in the formation of malfunctioning haemoglobin, giving rise to sickle cell anaemia.

Most mutations have no effect on phenotype, some have harmful effects, and some mutations can be beneficial to the organism. For example, mutations in houseflies which make them resistant to the pesticide DDT are an advantage when DDT is present. Even the sickle cell anaemia mutation can be advantageous. Heterozygotes who carry both sickle cell and normal alleles are more resistant to malaria. In countries where malaria is widespread, carrying the sickle cell allele gives a distinct advantage.

> **Extension**
>
> Approximately one in seven human mutations are caused by jumping genes. To find out more read **Extension 4.4** – Jumping genes. **X4.04S**

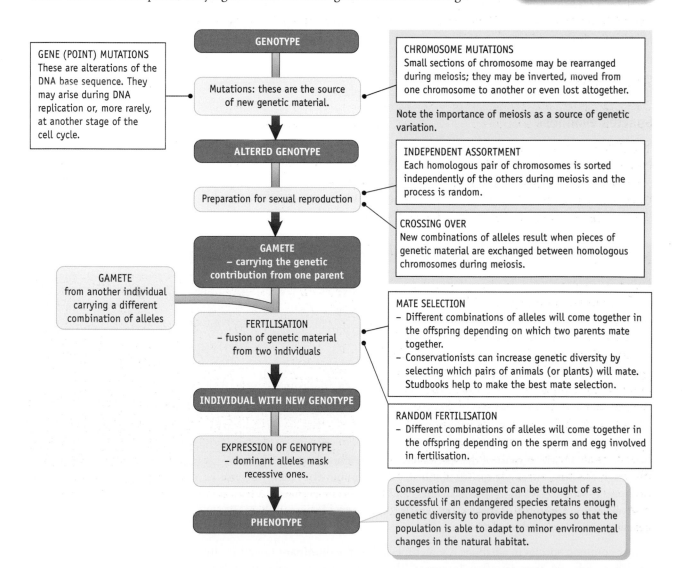

Figure 4.28 Sources of variation.

Genetic diversity may be visible

Sometimes genetic differences within species show up clearly in the external phenotypes. For example, wild arum has a variable number of spots on its leaves, and the spottiness increases in more northerly populations (Figure 4.29).

Much genetic variation has no visible effect on phenotype. It is made up of molecular differences which can only be detected using techniques such as gel electrophoresis of proteins; two alleles at a single locus produce slightly different protein products. However, these differences may be important in evolutionary terms.

Figure 4.29 Lots of natural variation reflects high genetic diversity within a species. Species with wide geographical ranges often show regional variations, sometimes with a gradual trend from north to south or east to west. Compare the differences between this wild arum and the one in Figure 4.15.

Measuring biodiversity

Species richness

The simplest way to measure biodiversity is to count the number of species present in a given habitat. This is called **species richness**. A habitat is a place where organisms live, such as a pond, a coral reef, a wood, a meadow, or a microhabitat such as the gap under a stone. A wood in southern England might have 25 butterfly species, but a similar area of forest in South America may have ten times that many. By this measure we can say that the South American forest has more biodiversity than the English wood. In practice it may only be possible to list some of the species present, for example vertebrates or vascular plants.

Species evenness

Species richness is a useful measure, but it takes no account of the population size of each species. For example, consider the hypothetical situation in Table 4.3. Wood A is clearly more diverse than woods B or C, because it has a greater number of species. But wood B is more diverse than wood C, though they both have five species. In wood B the levels of abundance of the individual species are more similar than in wood C.

Species	Wood A	Wood B	Wood C
blue tit	25	20	50
robin	20	17	12
wren	11	23	3
blackbird	9	22	27
willow warbler	17	0	0
great tit	7	18	8
blackcap	4	0	0
song thrush	7	0	0

Table 4.3 Abundance of bird species (%) in three woods.

A community in which most of the species have similar abundances is said to have high evenness; no single species dominates the community. A highly diverse community would have high species richness and high species evenness. A community with an identical number of species, but which is dominated by one of them, is generally considered to be less diverse.

A very common species in a habitat is sometimes called the **dominant** organism. In English woods, oak trees are often dominant. This is partly because, over many years, humans have selectively encouraged oak trees because they provide good timber. In some habitats, a small number of animal species such as pigeons and rats can become dominant at the expense of the rest, and become pests. In species-rich ecosystems, such as rainforests, species composition is usually more even.

Q4.17 Table 4.4 shows the tree composition in two imaginary British woods, one managed by humans and the other left in its natural state. Let us assume that in each wood one hundred trees were selected at random and identified. Compare the species richness and species evenness of the two woods and decide which has the greater biodiversity.

Activity

In **Activity 4.11** you can measure biodiversity for yourself. **A4.11S**

	Managed wood	Ancient wood
Oak	50	15
Ash	10	15
Sycamore	25	0
Field maple	2	15
Hazel	3	20
Small-leaved lime	0	8
Wild service	0	9
Common hawthorn	5	0
Midland hawthorn	0	10
Holly	2	8
Beech	3	0
Total	100	100

Table 4.4 Tree composition of two imaginary woods.

Extension

Many other methods are used to quantify biodiversity. These involve the calculation of some sort of biodiversity index, to give a numerical figure to indicate the biodiversity of a habitat. In **Extension 4.5** calculate Simpson's Indices for contrasting communities. **X4.05S**

Finding the hotspots

The world's biodiversity is not distributed evenly across the surface of the planet. In the last few decades, unexpectedly high levels of biodiversity have been discovered in unlikely places. For example, while coral reefs and rainforests are ranked as the richest ecosystems in the world, the **biodiversity hotspot** containing the largest proportion of the world's plant species is not, as you might expect, in the tropics. Instead, the plant hotspot is the Mediterranean Basin which contains one in ten of the Earth's plant species. More than half are endemic, that is found only in that area and nowhere else on Earth. Another example comes from ocean sea beds. These have long been regarded as very low diversity ecosystems. However, exploration of the muds of deep oceans has revealed a startling range of animals living in sediments previously thought to be lifeless.

The term 'biodiversity hotspot' was coined in 1988 by conservationist Norman Myers to describe areas of particularly high biodiversity. He identified 10 tropical forest regions containing between them 13% of all plant diversity in just 0.2% of the Earth's land area. Conservation International later adopted the idea as a way of focusing conservation effort on the most critical places. They have extended the list to include 25 regions collectively covering only about 2% of the Earth's land surface, yet containing more than 50% of all terrestrial species (Figure 4.30). These particular hotspots are identified in terms of plant diversity. To be regarded as a hotspot according to these criteria, a region must have a minimum of 1250 different plant species; crucially, it must also have 0.5% or more of total global plant diversity present as endemic species.

Weblink

You can find out more about the hotspots by visiting the Conservation International hotspots website and using their explorer map.

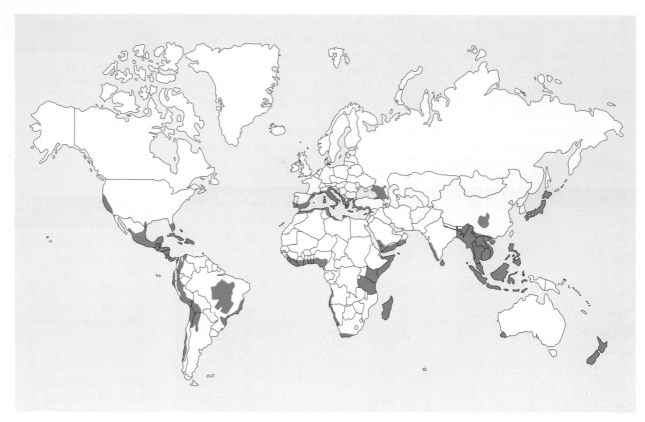

Figure 4.30 Biodiversity hotspots of the world.

Q4.18 Suggest why hotspots are identified only in terms of plant diversity.

Q4.19 Describe a plausible explanation for the distribution of hotspots shown on the map in Figure 4.30.

Measuring genetic diversity within a species

Genetic diversity can be measured directly or indirectly. The direct method is by DNA sequencing to determine the bases in a segment of DNA, and thus to determine which alleles are present. Alternatively, the DNA can be cut into fragments and then separated using electrophoresis. Different alleles can be identified because they produce fragments of different lengths. Electrophoresis is studied in detail in Topic 6.

The genes from a group of individuals are sampled, and the alleles present in this sample are recorded. The data are used to calculate an index which quantifies the genetic diversity. For example, the heterozygosity index is the proportion or percentage of genes which are present in the heterozygous form.

Conservation biologists may use genetic diversity indices to compare different populations when deciding which populations need to be conserved. If there is high genetic diversity within all the populations, then conservation effort can be focused on some populations without overall loss of genetic diversity.

Activity

In **Activity 4.12** you can compare the heterozygosity indices for natterjack toad populations. **A4.12S**

4.4 Making use of biodiversity

Look around you and notice how many of the things that you rely on come from natural biodiversity, plants in particular (Figure 4.31). Adapted to survival in their own environments, the structures and products made by plants are often what we depend on as resources for our own survival.

Weblink

The diversity of plant uses can be seen at the Plants for a Future website.

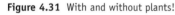

Figure 4.31 With and without plants!

Big and strong

Why do trees grow tall? Some plants lift their leaves on stalks or trunks so that they are above those of their competitors. If it can grow big enough, a tree can obtain massive amounts of photosynthetic energy, and sink roots deep into the ground to take in the maximum amount of water and nutrients. The Brazil nut tree can reach a height of over 50 m and grows up to 20 m above the forest canopy to obtain light (Figure 4.32). If attacked by fungi, viruses or animals, part of their structure may be lost, but the rest can live on.

This is a risky strategy for survival since it can take a long time to grow to maturity. The Brazil nut tree does not produce fruit until it is about 45 years old. Yet it works; plants can build significantly bigger structures than any land animal. Redwoods can grow to 90 m, and 20 m palm trees can survive 100 mph winds. Such large structures must be strong enough to hold up their own weight and withstand the enormous wind forces on them. The same applies to upright annual plants, though they don't grow so tall because they only have one growing season in which to reach maturity.

Figure 4.32 Brazil nut trees can tower 20m above the surrounding trees.

Building tall structures

All plants use three basic principles to build tall structures:

1 They produce strong cell walls out of cellulose, a polymer made from sugar molecules.

2 They build columns and tubes from specialised cells.

3 They stiffen these special cells with another polymer called **lignin**.

Trees add a ring of this stiffened (lignified) tissue each year, and building up wood in this way allows trees to grow taller. The structures produced are not only strong but also flexible, allowing trees to sway in high winds. We attempt to reproduce the properties of wood in composite materials, using more than one substance in combination. For example, concrete reinforced with steel is much stronger than either material alone.

 Did you know?

Reach for the sky

Like trees, tall buildings have to be strong enough so that they do not collapse under their own weight. Materials such as concrete provide strong walls that will bear this weight.

Tall buildings also have to deal with the horizontal force of wind. At their tops, most skyscrapers can move several feet in either direction, like a swaying tree, without damaging their structure. Any more movement would not only be uncomfortable for people inside, but would risk breaking up the concrete structures – these are brittle, unable to bend much without breaking.

Steel columns allow a degree of bending – they are much less brittle than concrete. However, a tall building made entirely from steel columns would eventually buckle and bend under its own weight. The solution is a steel framework combined with a concrete structure, providing both strength and stiffness (Figure 4.33). This allows the entire structure to move a little as one unit, like a tree.

Figure 4.33 Buildings have to withstand the same forces as tall trees. Architects copy plants when designing buildings, using concrete walls reinforced with steel instead of cellulose walls reinforced with lignin.

In order to appreciate how plant cells are highly specialised for their function you first need to study a typical plant cell.

Key biological principle: How are plant cells different from animal cells?

Figure 4.34 shows the ultrastructure of a generalised plant cell. Decide how it is different from the animal cell shown in Figure 3.8 (page 105).

There are two fundamental differences:

• The plant cell has a rigid cell wall.

• The plant cell contains chloroplasts.

Chloroplasts are the site of photosynthesis, where energy from the Sun is used to make storage molecules. Starch is found in storage vacuoles in the cytoplasm called amyloplasts. In addition to the cell wall and chloroplasts in plant cells, there is often a large central vacuole surrounded by a vacuolar membrane (tonoplast).

Activity

Visit the interactive cell in **Activity 4.13** to compare animal and plant cells. **A4.13S**

Checkpoint

4.4 Draw up a table or mind map comparing the typical ultrastructures of animal and plant cells.

Figure 4.34 A generalised plant cell.

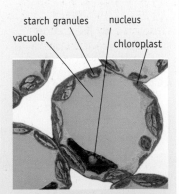

Figure 4.35 False colour electron micrograph showing parenchyma cells in the leaf contain many chloroplasts. Magnification x2000.

Parenchyma is a type of plant tissue found throughout the plant. The cells fill spaces between more specialised tissues, and may themselves have certain specialised functions. For example, in roots they may have a role in storage, and in leaves they contain chloroplasts and form the photosynthetic tissue (Figure 4.35).

CONTINUED ▶

> **Q4.20** The plant cell can be likened to a factory, with each of the organelles carrying out a special function. Look at the factory functions below and in each case decide which of the cell organelles performs an equivalent function:
>
> **a** control centre where the instructions are kept to tell the factory floor what to make
> **b** boiler room, harnessing energy to drive all the processes in the factory
> **c** packaging hall, wrapping up and collecting products to be exported from the factory
> **d** security screen, allowing in only those external raw materials which the factory needs
> **e** warehouse storage facility within the factory.

Cell walls – the secret of their strength

Cellulose

A plant's strength comes in part from the thin **cellulose** walls of plant cells and the 'glue' that holds them together.

> **Support**
>
> You can remind yourself about condensation reactions and –OH groups by looking at the Biochemistry support on the website.

A Two forms of glucose

α-glucose β-glucose

B Joining two β-glucose molecules

C Glucose chain – cellulose

Figure 4.36 A Cellulose is made up of β-glucoses joined together. Starch is composed of α-glucoses.
B Formation of a 1,4 glycosidic bond between the two β-glucose molecules is only possible if one molecule is rotated through 180°. **C** A molecule of cellulose. Notice how the –OH groups project from both sides of the molecule due to the arrangement of the glucose molecules. Each alternate glucose is inverted to allow the 1,4 glycosidic bond to form.

Cellulose is a polysaccharide.

It is a polymer of glucose, but the glucose it is made from is slightly different from that which forms starch. In Figure 4.36 the two forms of glucose are shown alongside each other. Can you spot the difference? You should notice that the –OH (hydroxyl) groups on the first carbon atoms are on opposite sides.

Cellulose is made up of β-glucose units. A condensation reaction between the –OH group on the first carbon of one glucose and the –OH on the fourth carbon of the adjacent glucose links the two glucose molecules. A 1,4 glycosidic bond forms. In cellulose, all the glycosidic bonds are 1,4; there are none of the 1,6 glycosidic bonds that occur in starch. Because of this, cellulose is a long unbranched molecule.

Q4.21 Look at Figure 4.36B. Describe the relative positions the two β-glucose molecules must have for a condensation reaction to take place between them.

Q4.22 Explain why the bond between each pair of glucose molecules in cellulose is called a 1,4 (glycosidic) bond.

Microfibrils are bundles of cellulose molecules

Each cellulose chain typically contains between 1000 and 10000 glucose units. Unlike an amylose molecule that winds into a spiral (Figure 1.35, page 34), the cellulose molecules remain as straight chains. Hydrogen bonds form between the –OH groups in neighbouring cellulose chains, forming bundles called **microfibrils** (Figure 4.38). Individually, the hydrogen bonds are relatively weak compared with the glycosidic bonds, but together the large number of hydrogen bonds in the microfibril produces a strong structure.

Figure 4.38 In cellulose, neighbouring chains of glucose molecules are linked by hydrogen bonds to form microfibrils.

If you look more closely at a plant cell wall (Figure 4.39), you can see that it is formed of microfibrils. These microfibrils are bundles of about 60–70 cellulose molecules. The microfibrils are wound in a helical arrangement around the cell and stuck together with a polysaccharide glue. Figure 4.39 shows how successive layers of the microfibrils are laid down at angles to one another, forming a composite structure.

The glue that holds together the microfibrils is composed of short, branched polysaccharides known as **hemicelluloses** and **pectins**. These short polysaccharides bind both to the surface of the cellulose and to each other, and hold the cellulose microfibrils together.

Pectins are also an important component of the middle lamella – the region found between the cell walls of adjacent cells. The pectins act as cement and hold the cells together.

The arrangement of the cellulose microfibrils within a matrix of hemicelluloses and pectins makes the cell wall very strong – rather like steel-reinforced rubber tyres. In this analogy, the hemicellulose matrix is the rubber, and the cellulose microfibrils are the reinforcing steel cables. This makes a strong but pliable structure. The microfibrils are laid down at different angles, which makes the wall strong and flexible.

Figure 4.39 In the cell wall, layers of cellulose microfibrils are laid down at different angles.

? Did you know?

Cross-linking

Looking at a pair of nylon tights under the microscope shows that there are tight chains running around the leg of the tights. These chains are joined more loosely vertically, so there is more 'give' lengthways. Net bags and elastic tubular bandages also show this uneven stretch. It is cross-links between the units that determine the amount of movement, and so the strength in any direction. Cross-linkages also strengthen scaffolding, electricity pylons and steel girders (Figure 4.37).

Figure 4.37 This structure relies on cross-linking to provide the structure with strength.

⚙ Activity

In **Activity 4.14** you can use the Biochemistry support on the website to see how β-glucoses join to form cellulose molecules, and also to compare the structures of starch and cellulose. **A4.14S**

Checkpoint

4.5 Look back at the structure of starch in Topic 1. Draw up a table comparing the structures and functions of starch and cellulose.

Pectins, jams and fruit juices

When fruits ripen they usually become soft (and eventually squashy) as the pectins holding the cells together are broken down by enzymes. The pectins become soluble, allowing the cells to shift when you squeeze the fruit.

In the fruit juice industry pectins are a nuisance. They thicken the juice, making it difficult to extract from the fruit, and slowing down filtration. Enzymes called pectinases are used commercially to break down the pectins, increasing yield and giving juices such as apple juice their clarity. Pectinases are also used to make baby food purées, and to peel the citrus fruits that come in tins.

Conversely, when making jam, pectins are a good thing – the more pectins there are, the better the jam sets, and the less fruit you have to use per jar. But if the fruit is too ripe the natural pectins will have broken down, and the jam will remain runny unless more pectins are added.

Crossing the cell wall

Cell walls do not separate plant cells completely. Narrow fluid-filled channels, called **plasmodesmata**, cross the cell walls, making the cytoplasm of one cell continuous with the cytoplasm of the next. Cell walls are also fully permeable to water and solutes. Can you identify the plasmodesmata in the cell wall shown in Figure 4.40? (Look at the photograph and decide before reading the caption.)

At some places the cell wall is thin because only the first layer of cellulose is deposited. The result is a **pit** in the cell wall. Plasmodesmata are often located in these pits, aiding the movement of substances between cells.

Figure 4.40 Photo of a section through a cell wall showing the channels between adjacent cells. The plasmodesmata are visible as parallel lines crossing the cell walls.

Tubes for transport and strength

To build a tall plant such as a tree, some of the cells within the stem must be stiffened to provide mechanical support. At the same time, some cells must allow water and minerals (inorganic ions) to pass from the roots to the leaves.

There are two specialised types of cell of particular importance in fulfilling these functions. These are:

- xylem vessels – these form tubes for transport, and their stiffened cell walls help support the plant
- sclerenchyma fibres – columns of these cells with their stiffened cell walls also provide support.

Where in the stem are these specialised cells?

There are three basic types of tissue found within plants. Figure 4.41 shows the general location of these three types: dermal tissue (epidermis), vascular tissue and ground tissue.

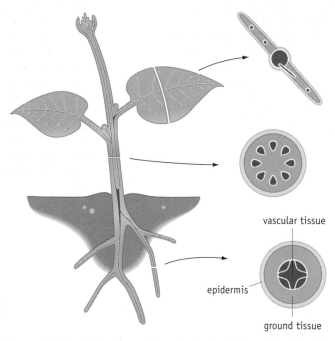

Figure 4.41 Where are the three basic types of tissue in the plant? The epidermis is a single layer of cells covering the entire outside of the plant. The vascular tissue, involved in transport, is surrounded by the ground tissue which contains cells specialised for photosynthesis, storage and support.

Figure 4.42 shows the location of the **vascular tissue** within a cross-section of a stem from a dicotyledon (Figure 4.43). Notice how each **vascular bundle** contains **xylem vessels** and **phloem sieve tubes**. On the outside of the bundle are **sclerenchyma fibres**. In a young dicotyledon, the vascular tissue is in bundles towards the outside of the stem. In trees and shrubs these separate bundles merge to form a continuous ring as the plant ages. The xylem vessels carry water and inorganic ions up through the stem. The phloem transports sugars made by photosynthesis in the leaves up and down the plant.

Activity

In **Activity 4.15** you can examine plant stems and locate the different tissue types yourself. **A4.15S**

Q4.23 Look at Figure 4.42C. Identify the locations of:

a   a vascular bundle

b   xylem vessels

c   sclerenchyma fibres.

A

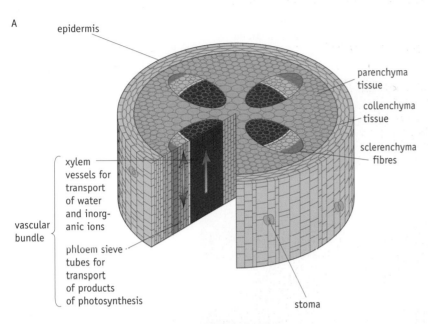

epidermis

parenchyma tissue

collenchyma tissue

sclerenchyma fibres

xylem vessels for transport of water and inorganic ions

vascular bundle

phloem sieve tubes for transport of products of photosynthesis

stoma

Figure 4.42 A The general arrangement of vascular tissue within the stem of a dicotyledon. The arrows represent the direction of flow through the xylem and phloem. **B** Coloured scanning electron micrograph of a vascular bundle of a buttercup, *Ranunculus repens*. The phloem sieve tubes are orange/yellow with the sclerenchyma fibres on their left. Magnification ×135. **C** A photomicrograph of a transverse section of a sunflower, *Helianthus*, stem.

Figure 4.43 Flowering plants, the angiosperms, can be divided into two classes depending on the number of seed leaves they have. Monocotyledon embryos have one seed leaf, their leaves have parallel veins and they rarely grow large because they cannot produce true wood. Dicotyledon embryos have two seed leaves and the veins on their leaves form a network pattern. They can produce wood and so include trees and shrubs. The grass on the left is a monocotyledon. The clover is a dicotyledon.

Strong tubes

Tubes are useful things. They can carry materials, and because they are hollow they are light. But tubes have a tendency to collapse in on themselves when put under tension. So they need to be reinforced if they are not already made of a strong material (Figure 4.44).

Figure 4.44 This large-scale vacuum tube uses spiral thickening to hold it open.

Xylem vessels for transport

Figure 4.42A shows a cross-section of a plant stem. The xylem vessels are made up of large cells with thick cell walls. They form a column of cells acting as tubes for the transport of water and mineral ions. In order to transport water, the cell walls have to be waterproofed. The plant does this by producing another polymer – **lignin**. This polymer impregnates the cellulose cell wall and, as the cells become lignified, the entry of water and solutes into them is restricted. At about the same time, the tonoplast breaks down, and there is **autolysis** of the cell contents. During autolysis, the cell organelles, cytoplasm and cell surface membrane are broken down by the action of enzymes and are lost, leaving an empty tube.

The detailed structure of xylem vessels is shown in Figure 4.45. The end walls between the cells of the columns are lost or become highly perforated. Long tubes form as a result of this process, and they are continuous from the roots of the plant to the leaves. The cellulose microfibrils and lignin in the cell walls of the xylem vessels give the tubes great strength.

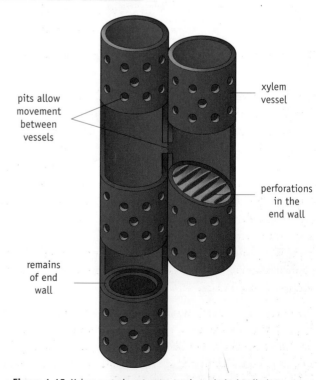

pits allow movement between vessels

xylem vessel

perforations in the end wall

remains of end wall

Figure 4.45 Xylem vessels cut open to show their detailed structure.

Q4.24 Explain how disruption of the tonoplast might result in the breakdown of the cell contents.

How is water transported through xylem vessels?

Xylem vessels are effectively fluid-filled tubes through which water moves upwards from the roots to the shoots. How is the water moved through these dead cells?

Water evaporates from all surfaces of the plant, mostly from the large surface area of the leaves. The majority of the evaporation occurs from the surfaces of the cells that line the substomatal cavities in the leaves, as shown in Figure 4.46. Water diffuses out through the stomata down a **diffusion gradient**. Water evaporating from the plant in this way is known as **transpiration**. The water that leaves a plant leaf by transpiration is replaced by water absorbed through the roots. In the tallest trees water may move up to 100 m through the plant.

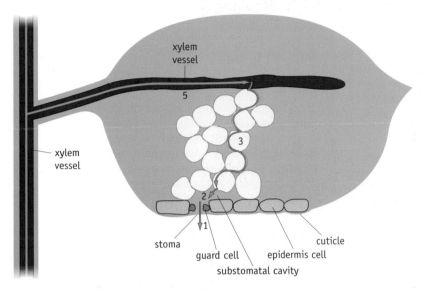

1 Water vapour diffuses out through the stoma down a diffusion gradient.

2 Water evaporates from the surfaces of cells lining the substomatal cavity.

3 Water replaced by means of capillary action within the cell walls.

4 Water is drawn out of the xylem.

5 A continuous column of water is drawn up through the xylem.

Figure 4.46 Water moves up through the xylem due to the evaporation of water from the surfaces of cells within the leaf.

It is the evaporation of water from the cells in the substomatal cavities of the leaves that provides the force needed to draw water up a plant. The minute channels between the cellulose microfibrils in the cell walls act as tiny capillaries. Capillaries have the ability to draw water up them by capillary action, which is caused by surface tension. As water evaporates from the surfaces of the cell walls within the leaf, it is replaced by means of capillary action within the cell walls. This draws more water to the cell surfaces. The thousands of minute capillaries in the cell walls inside the leaves produce a massive pull on the water behind them. This pull is thought to be sufficient to draw water from the xylem vessels, and up the whole stem of the plant, even in large trees. Water moves up the xylem vessels and through the cell walls within the leaf in a continuous stream. This stream of water passing through the plant is known as the **transpiration stream**.

The energy for moving water through a plant comes from the Sun, which heats and evaporates water from the cells lining the substomatal cavity. Water is pulled up from the top of a plant rather like pulling on a piece of string. The water is under tension as it is pulled up. Xylem vessels would collapse inwards, like a strongly sucked straw, but for the reinforcing lignification of their walls. As in a straw, the water column does not break when it is pulled. This is because of the cohesive forces between water molecules, which are a result of hydrogen bonding – water in narrow tubes sticks together very strongly. Because the water is linked by cohesion and is pulled up under tension, this concept of water movement is known as the **cohesion-tension theory**.

Activity

The interactive tutorial in **Activity 4.16** should help you understand the function of xylem in the transport of water through the stem. Use the Biochemistry support on the website to remind yourself about water and hydrogen bonding. **A4.16S**

Key biological principle: The importance of water to plants

Topic 1 described water as a polar molecule. This polarity results in the formation of hydrogen bonds between water molecules. Polarity is the cause of many of the properties of water, including its being liquid at room temperature (see page 8).

Cohesion and surface tension

Hydrogen bonding between water molecules results in strong cohesive forces between water molecules that keep the water together as a continuous column in xylem vessels (Figure 4.47).

hydrogen bonds between water molecules – cohesion

hydrogen bonds between water molecules and the cell wall – adhesion

Figure 4.47 A sketch to show the forces of cohesion in a xylem vessel.

Surface tension at water surfaces is also partly caused by these cohesive forces between water molecules, which cause the surface layer of water to contract. This is useful for some small aquatic organisms, such as duckweed and pond skaters, which can be supported on this surface film.

Solvent properties

The solvent properties of water described in Topic 1 mean that dissolved substances can be transported around plants through the xylem and phloem. Once in cells, the dissolved chemicals move freely around in an aqueous environment and can react, often with water itself being involved in the reactions, for example in hydrolysis and condensation reactions. The synthesis of sugars from water and carbon dioxide in photosynthesis is important to plants, and to all organisms which depend on plants for their food supply. The details of photosynthesis and its role in the carbon cycle will be studied in Topic 5.

Thermal properties

Q4.25 Recall from Topic 1 what specific heat capacity is and explain why it is very high for water.

You should remember that a large input of energy causes only a small increase in temperature, so water warms up and cools down slowly. This is extremely useful for organisms, helping them to avoid rapid changes in their internal temperatures, and enabling many of them to maintain a fairly steady temperature even when the temperature of their surroundings varies considerably.

Density and freezing properties

Unlike most liquids, water expands as it freezes. As liquid water cools, the molecules slow down, enabling the maximum number of hydrogen bonds to form between the water molecules. These hydrogen bonds hold the water molecules further apart than in liquid water, making ice less dense than liquid water. So ice floats, enabling organisms to survive in liquid water under the ice in frozen oceans, ponds and lakes.

What else do xylem vessels transport?

The movement of water through the xylem provides a **mass flow system** for the transport of inorganic ions. These are absorbed into the roots and are required throughout the plant.

Q4.26 There is a very low concentration of inorganic ions in the soil solution surrounding roots. Using what you know about transport across cell surface membranes, suggest how the inorganic ions are transported into the root cells.

Activity

In **Activity 4.17** you can investigate plant mineral deficiencies practically. **A4.17S**

Nitrate ions are needed by plants in order to make amino acids. Amino acids contain one or more nitrogen atoms. Unlike us, plants make all their own amino acids from scratch using inorganic materials. By a sequence of enzyme-controlled reactions, the nitrogen from nitrate ions (transported in the xylem) is combined with organic molecules from photosynthesis to make all 20 amino acids. Plants cannot grow without nitrate ions; like all organisms, their cell cytoplasm is made largely of proteins, built by joining amino acids together (see Topic 2 page 62). Some other important biological molecules found in plants also contain nitrogen atoms, notably chlorophyll, nucleic acids, ATP and some plant growth substances.

If inorganic ions are not absorbed in sufficient amounts, the plant will show deficiency symptoms. For example, if the plant lacks magnesium it is unable to make chlorophyll and the older leaves become yellow (yellow leaves may also be a sign of nitrogen deficiency) (Figure 4.48). A lack of calcium causes stunted growth due to the role of calcium ions in the structure of the cell wall and in the permeability of the cell membrane.

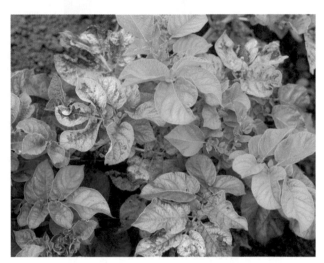

Figure 4.48 This potato plant is lacking in magnesium.

Xylem and sclerenchyma for support

Lignin not only waterproofs the cell walls, it also makes them much stiffer and gives the plant much greater tensile strength. Instead of forming a uniform layer on the inside of the xylem cell walls, lignin is often laid down in spirals or in rings as shown in Figure 4.49.

Xylem vessels are not the only cells that become impregnated with lignin. The sclerenchyma fibres associated with vascular bundles in the stem and leaves also have lignin deposited in their cell walls. 'Sclerenchyma' comes from the Greek *scleros* meaning 'hard'. As with xylem vessels, the sclerenchyma fibres die once lignified, leaving hollow fibres (Figure 4.50). The strength of these fibres varies in different plant species depending on the length of the fibres and the degree of lignification.

The taller a plant needs to grow, the greater the proportion of its stem that becomes lignified. In a tree, this is the majority of the trunk: the living parts are towards the surface (under the bark) and grow new layers each year. These layers are the annual rings. In annual plants the lignification is confined to the vascular bundles. The plant stem relies on tightly packed, fully turgid parenchyma cells to maintain its shape and keep it erect. A **turgid** cell is one that is completely full, with its cell contents pressing out on the cell wall. Turgor supports the leaves of all plants including those of trees. If a cell loses water, turgor is lost. If a high proportion of a plant's cells lose their turgidity, the plant wilts.

Figure 4.49 The lignin thickening in xylem vessels can take different forms as shown in this photomicrograph.

thick wall of lignin

chisel-shaped end of fibre

Figure 4.50 A Several sclerenchyma fibres. **B** Part of two fibres shown at a higher magnification.

Useful plant fibres

Humans use plants in many ways. People have been using plant fibres for thousands of years in order to make products such as clothing, rope, floor coverings, paper and many more. Plant fibres can be used in these ways because they are:

- long and thin
- flexible
- strong.

How do we extract fibres from plants?

To obtain fibres we must take the plant apart. This can be done mechanically by pulling out the fibres or by digesting the surrounding tissue. Fortunately for us, cellulose – and particularly cellulose combined with lignin – is very resistant to chemical and enzymic degradation, whilst the polysaccharides that hold the fibres together can be dissolved away.

The more lignin there is present, the harder it is to separate fibres. So to produce fibre pulp from trees, caustic alkali is required. For flax (Figure 4.51) and other suitable plants a milder treatment is used, and in some traditional processes the stems are piled in heaps, allowing bacteria and fungi to do the work. This process, and its more modern chemical and enzymic equivalents, is called 'retting'.

Figure 4.51 Scanning electron micrograph of fibres from a flax plant (*Linum usitatissimum*) used in linen fabric.

Q4.27 What do you think makes plant fibres a success commercially?

 Checkpoint

4.6 Look back over all the information on xylem vessels and sclerenchyma fibres. Then list the physical properties that enable them to be used for human benefit.

 Activity

In **Activity 4.18** you can have a go at extracting fibres and then testing their tensile strength. **A4.18S**

Not just textiles ...

Fibres have many other uses in addition to the textile industry. For example, mats of fibres are used to absorb heavy metals and also hydrocarbons from polluted water (Figure 4.52).

Plant fibres can also be added to other materials to form biocomposites. For example, when oilseed rape (also known as rapeseed) fibres are mixed with plastic, the resulting material is stronger than plastic alone. These biocomposites are also renewable, more biodegradable and can be easier and safer to handle than composites containing artificial fibres.

Researchers at Warwick University have built a car from biomass. Their Eco One sports car has tyres made from potatoes, brake pads made of ground cashew shells and a body built from hemp and rapeseed. Only the car's steering-wheel, seat, electrics and chassis are made from conventional materials. The car runs on a special biofuel made entirely from sugar beet and fermented wheat. It is not the first time such a car has been produced: Henry Ford produced a car made entirely from hemp in 1941.

Figure 4.52 Oil pollution control. Natural fibres absorb hydrocarbon pollution from water.

Chemical defences against attack

A plant can't move much, it can't run and it can't hide, so it is an easy target for animals, bacteria and fungi. The agouti eats most of the Brazil nut tree's nuts, but the tree benefits as it helps disperse its seeds. However, many plants have adaptations that provide chemical defences to repel and even kill animals that feed on them. One strategy is to produce a chemical that is distasteful or even toxic. If the animal takes a bite and the taste is offensive (usually bitter, Figure 4.53), the animal is deterred from feeding further. If its chemicals kill the predator, the plant will avoid future attacks.

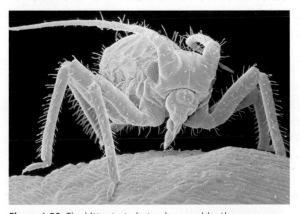

Figure 4.53 The bitter taste in tea is caused by the presence of tannins in the leaves. These also have the effect of 'gumming up' the mouthparts of aphids and so protect the tea plant from this form of insect attack.

Pyrethrum – natural insecticides

Pyrethrum is a kind of chrysanthemum and a member of the daisy family (Figure 4.54). Pyrethrum has the remarkable property of having no known pests or diseases. It contains chemicals which are extremely toxic to insects. Pyrethrum is grown commercially in Kenya, where flowers are picked, dried in the sun, then bagged and transported to factories. Here they are crushed into powder and further refined to extract the active ingredients: various pyrethrins.

Pyrethrins are used in many insecticide recipes in their native home, Kenya. For example, they're used in sprays for fruit and flowers, flea treatments for cats and dogs, and the reduction of tick infestation in cattle.

Fortunately, these compounds are much less toxic to vertebrates than to insects, but they are not entirely harmless and must be handled with care. Pyrethrins have the advantage that they are unstable after spraying and rapidly decompose to harmless residues.

Figure 4.54 *Tanacetum coccinium*, the pyrethrum plant.

One alternative to using pyrethrins is to use the plants themselves. Organic gardeners use chrysanthemums and marigolds as 'companion' plants – planted amongst vegetables, they act as a natural insect repellent.

You might try growing some chrysanthemums and testing them on flies.

Natural antibacterials

Plants sometimes store toxic compounds in hairs on the surfaces of their leaves. This is very obvious in the stinging nettle, which most of us avoid, but not so apparent in mint, which produces the chemicals menthol and carvone. These chemicals are toxic to microbes and some insects, but attractive to us as flavouring in foods or tea. Even humans experience the numbing effect of mint (whilst brushing your teeth, for example) but to microorganisms this can be lethal.

Garlic extracts have been found to destroy bacteria such as *Campylobacter* and *Helicobacter* which cause intestinal infections. This is potentially important as some strains of the bacteria are now resistant to widely used antibiotics such as penicillin. The active ingredient in garlic is allicin; this is known to interfere with lipid synthesis and RNA production. Allicin is only produced when the plant is cut or damaged. Its inactive precursor, alliin, is converted into the active form by the enzyme alliinase.

Some studies have shown that certain parts of a plant tend to have greater antibacterial properties than the rest. These are typically the seed coat, fruit coat, bulb and roots.

Q4.28 Suggest why the parts of the plant listed above need to produce the most antibacterial chemicals.

Activity

You can investigate the antibacterial properties of plants in **Activity 4.19. A4.19S**

Medicines from plants

Not surprisingly, many plants contain poisons, or produce them rapidly as a response to wounding. However, 'poison' is a relative term and relates to the dose necessary to cause harm to an organism. Clearly, if a chemical can kill pathogenic (disease-causing) microbes or malignant cancer cells at a dose level which leaves humans alive, then this 'poison' is a likely medicine. An enormous number of medicines are derived from chemicals originally discovered in plants.

The World Health Organization estimates that 75–80% of the world's population uses extracts from plants as medicines. Many of the common medicines we use were originally plant-derived, such as aspirin (salicylic acid derived from an extract of willow bark), and morphine and codeine (both derived from opium poppies).

Digitalis and drug development

Foxgloves and dropsy

Foxglove leaves are poisonous when eaten by humans and other animals. They have a strong, bitter taste which serves as a warning. The symptoms of poisoning are dizziness, vomiting, hallucinations and heart failure caused by an irregular heartbeat. But it is the effect on the heartbeat that made it a traditional folk remedy when used in moderation.

Figure 4.55 The foxglove, *Digitalis purpurea*, has been used in medicines for hundreds of years.

The foxglove (Figure 4.55) was known for centuries to have medicinal qualities and in particular was used to treat a condition known as dropsy. Dropsy, now called oedema, happens when fluid accumulates in the body tissues (see Topic 1 page 27). This process is painful and can cause a slow death. Oedema is usually caused by heart or kidney problems. A fast and irregular heartbeat is one of the signs.

Because the blood pressure is raised, tissue fluid fails to return to the capillaries and accumulates in the patient's feet, legs and organs, causing them to swell up. Eventually the patient may drown as fluid fills up their lungs. So it was not surprising that a herb known to relieve this serious condition was used as a remedy. However, it was not until a country doctor called William Withering (Figure 4.56) published *A Treatise on the Foxglove* in 1775 that it became an accepted form of medicine.

Withering had heard of the foxglove's curative properties for dropsy, but his attention was focused when he met Mrs Hutton, a 'wise woman' who was showing signs of the disease. She assured him that she would be all right after she'd had a cup of her special tea. Imagine Withering's surprise when he visited her again and found that she had recovered.

Talking to Mrs Hutton, Withering found out that foxglove was amongst the twenty or so herbs used in the potion. He suspected that it was something in the foxglove which was the active ingredient. Later Mrs Hutton sold her recipe to Withering who began to investigate the plant further. One of his first patients was a brewer suffering from swollen limbs and an irregular heartbeat. After a few doses of Withering's 'digitalis soup' he became healthy and his pulse became 'more full and regular'. Unfortunately his next patient, an old woman, nearly died from the treatment so Withering gave up his investigations.

Figure 4.56 William Withering.

Getting the dose right

After moving to Birmingham General Hospital, which received many patients suffering from dropsy, Withering was persuaded to renew his investigations. Studying

163 patients, he discovered and recorded side effects of digitalis. These included nausea, vomiting, diarrhoea and a rather strange green/yellow vision. A sign that the patient was recovering was the production of a large quantity of urine.

Withering realised that getting the dose right for the patient was of vital importance. He applied a standard procedure to discover the correct dosage for each patient. He slowly increased the dose until the patient started to have diarrhoea and vomiting, and then reduced it slightly; this would be the most effective dose. Withering meticulously recorded all his results, and after ten years wrote his book about the medicinal properties of the foxglove. This helped change the face of medical practice forever. We now know that the active ingredient in the foxglove is the chemical called digitalin.

> **Activity**
>
> You make a comparison between William Withering's approach to drug development and that of the drug companies in **Activity 4.20. A4.20S**

Q4.29 List the sequence of steps that Withering took when testing his drug.

Drug testing today

Today a potential new drug must pass a series of tests if it is to be developed into a new product. It has to be proven to be effective, safe and capable of making a profit. It can typically take 10–12 years and cost over US$1 billion to develop a drug.

Potential substances are analysed and the active ingredient (the drug compound which may bring about a cure) is identified and copied so that it can be manufactured synthetically. Slight variations of the chemical structure are made just in case they might have a better effect. (Nowadays many new drugs are 'discovered' through computer modelling of their chemical structures.)

A series of trials of the compound now begins. There are five stages in all. The first is pre-clinical testing, followed by three phases of **clinical trials** and finally after licensing trials. Table 4.5 describes these five stages.

Pre-clinical testing	Animal studies and laboratory studies on isolated cells and tissue cultures assess safety and determine whether the compound is effective against the target disease. These tests can take several years to complete. Thousands of chemicals go through pre-clinical testing, but only a handful are ever approved for clinical trials on humans. Animal trials form the basis of applications for clinical trials on humans, which are authorised by the Medicines and Healthcare products Regulatory Agency (MHRA), an independent body of scientists, doctors and members of the public.
Clinical trials – phase I	A small group of volunteers are told about the drug and given different doses. These volunteers are normally healthy, but there are circumstances where drugs are first tested on sick patients. The trial confirms whether or not the compound is being absorbed, distributed, metabolised and excreted by the body in the way predicted by the laboratory tests. The effects of different doses are monitored. In the UK a review of the data collected is made by the UK MHRA.
Clinical trials – phase II	Small groups of volunteer patients (100–300 people with the disease) are treated to look at the drug's effectiveness. If the results are promising, phase III trials are set up.
Clinical trials – phase III	A large group of patients (1000–3000 people) is selected and divided randomly into two groups. One group is given the compound being investigated. The second is given an inactive 'dummy' compound known as a **placebo**. If there is an existing treatment for the disease, the standard treatment is given rather than a placebo. It is important that neither the patients nor the doctors know who is having the compound under investigation and who is having the placebo or standard treatment. This is known as a **double-blind randomised controlled trial**, and is considered the 'gold standard' of valid testing. If the compound being investigated is effective, then the results will show a statistically significant improvement in the patients receiving the treatment compared with patients given the placebo or standard treatment. The tests also look for any adverse reactions in the patients. The way is now open to license the compound as a drug, after which it can be marketed.
After licensing	Trials continue to collect data on the effectiveness and safety of a new drug after the drug has been licensed.

Table 4.5 Stages in drug testing.

Q4.30   **a**   Explain why the review team in the phase I trials needs to be independent.

     b   If phase II is a success, why is phase III needed?

     c   Is it ethical to give some patients a placebo, when it is known that the compound is likely to have a beneficial effect?

     d   Sometimes patients on a placebo will show an improvement in their condition. Suggest why this may be so.

Seeds for survival

Having successfully survived the onslaught of all sorts of herbivores, the plant's next challenge is to make sure enough of the next generation also survive. Flowering plants have achieved this by packaging a miniature plant in a protective coat with its own food supply; we call them seeds. Inside the seed the embryo remains dormant until conditions are suitable for restarting growth (Figure 4.57).

Seeds are vital to the survival of a plant. They are adapted to ensure they:

* protect the embryo
* aid dispersal
* provide nutrition for the new plant.

What's in a seed?

In flowering plants the ovule is fertilised by the nucleus from a pollen grain and develops into the seed (see Topic 3 page 112).

The outer layers of the ovule become lignified forming a tough seed coat which protects the embryo within the seed. The surrounding ovary develops into the fruit (Figure 3.17) which often has an important role in seed dispersal.

In some species the stored food in the seed remains outside the embryo in storage tissue called endosperm, as shown in Figure 4.58. This is common in monocotyledons, for example cereals. Seeds of this type are called endospermic. In many dicotyledons the embryo absorbs the stored nutrients from the endosperm and the food is stored in the seed leaves (cotyledons) which swell to fill the seed (Figure 4.59). In some seeds, including Brazil nuts, there are no apparent cotyledons and the food is stored in the hypocotyls, the developing stalk.

Figure 4.57 The seed ensures that the embryo it contains has the best chance of germinating and establishing a new plant.

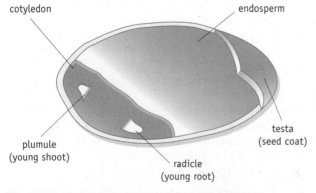

Figure 4.58 The internal structure of an endospermic seed.

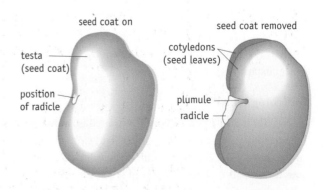

Figure 4.59 Opening up a broad bean reveals the cotyledons (seed leaves), plumule (young shoot) and radicle (young root) that make up the embryo.

Seeds come in all sizes and shapes, most of which are appropriate for wide dispersal. This helps offspring to avoid competing with their parent plant or with each other and it also lets plants colonise new habitats. Most species are adapted to take advantage of just one method of dispersal – wind, animal, water or self-dispersal.

Q4.31 Look at this list of adaptive features and decide which method of dispersal is being exploited in each case: wind, water, animals or self-dispersal.

   a   extremely small and light seeds, e.g. orchids

   b   seeds with wings, parachutes or similar structures, e.g. sycamore, dandelion and willow herb

   c   dehiscence – explosive rupture of a seed pod, e.g. peas and laburnium

   d   hooked fruits, e.g. burdock and buttercup

   e   fleshy fruits, e.g. blackberry

   f   fibrous seed coats containing lots of air, e.g. water lily and coconut.

When conditions are suitable and any dormancy has been broken, the seed takes in water through a small pore in the seed coat. Absorbing water triggers metabolic changes in the seed. Production of plant growth substances is switched on and these cause the secretion of enzymes that mobilise the stored food reserves. Maltase and amylase break the starch down into glucose which is converted to sucrose for transport to the radicle and plumule. Proteases break down the proteins in the food store to give amino acids and lipases break down the stored lipids to give glycerol and fatty acids.

What can we do with starch from seeds?

Although cereals are grown mostly for human food and animal feed, their carbohydrate polymers and oils also have major industrial uses.

Starch is easy to extract from plants. This is because it is in granules which do not dissolve in water, but can be washed out. In wheat, the protein remaining is gluten, which is a rubbery mass. The elastic properties of gluten are vital in bread-making, but are not much used in industry.

We all eat starch, but it has many other uses. Starch is found in a wide range of products including adhesives, paints, textiles, plaster, insulating material and toiletries such as conditioners, mousses, sun screens and anti-perspirants.

Thickening

When starch granules are heated in water they suddenly swell, absorb water, and thicken the liquid. This is 'gelatinisation', and this thickening process is the basis of both custard and wallpaper paste.

Stiffening fabrics

The stiffening of cloth or paper by starch is known as 'sizing', and enormous amounts of starch are used in paper coatings and cloth treatment. A starch mixture applied to the surface is gelatinised and then cooled, allowing bonds to form between the starch molecules. The addition of water reverses the stiffening. This reversal is called 'plasticisation' (the material becomes flexible again) and water is the plasticiser. This reversible behaviour is very useful, since by the addition of a little water and heat, the shape can be changed many times.

Super-absorbents

If starch is chemically cross-linked before it is gelatinised then particles are formed which can be dried. When rehydrated these particles can take up large amounts of water. Such cross-linked starch can now be found as super-absorbents in nappies.

Starch foam

The temperature at which starch gelatinises depends on the amount of water present. At water contents of less than 10% the gelatinisation occurs at much higher temperatures, above the boiling point of water. If the pressure is raised at the same time, the starch forms a plastic mass; if the pressure is suddenly released, for example when the seed coat ruptures during cooking, then steam forms and the starch 'puffs' into an expanded structure.

Activity

Have a go at popping corn yourself in **Activity 4.21** to demonstrate what happens to gelatinised starch when pressure is released. **A4.21S**

Puffed wheat breakfast cereals, corn snacks and starch-based foam packaging (Figure 4.60) are all made using this physical process, though corn snacks and starch-based foam are made in an extruder where a machine applies the pressure, rather than the seed coat. As hot starch leaves through a small exit hole the pressure that has built up inside is released, causing the starch to expand into foam as super-heated water turns to steam.

Starch-based packaging can now be used instead of polystyrene, polyethylene or other oil-based plastics. Plant-based plastics have also been developed which are made by fermentation of sugars from wheat, sugar beet, potatoes or agricultural waste. This means that supermarket food packaging such as a foam tray covered in plastic film can now be made from starch.

Figure 4.60 Foam packaging is commonly used to protect goods such as computers and telephones from knocks during transit and, to some extent, from large temperature fluctuations.

What can we do with vegetable oils?

Seeds are a rich source of oils which we regularly use in cooking. Oils are not only used for food though – many other industrial uses have been developed.

Fuels

There is nothing new about using vegetable oils instead of petroleum-based products for motor vehicles. Castrol, the engine oil, was originally derived from the castor bean. Dr Rudolf Diesel developed the first diesel engine to run on vegetable oil, and Diesel's engine was demonstrated at the 1900 World Exhibition in Paris using peanut oil. Biodiesel produced today can be used in unmodified diesel engines alternating with petroleum diesel. It produces less sulphur dioxide than diesel, and less carbon dioxide when you take into account the carbon dioxide used by the plants grown to produce it.

Across the UK there is a growing market for biodiesel. It is possible to buy biodiesel made from waste vegetable cooking oil, or from oil crops such as rapeseed. It is available either as 100% biodiesel, or as a blend with fossil fuel diesel, for use in standard diesel engines.

Sustainability

The use of oil-based plastics and fuels is not sustainable for several reasons.

- Burning oil-based fossil fuels releases carbon dioxide into the atmosphere, contributing to global warming (see A2 Topic 5).
- Oil reserves will eventually run out.
- Plastics generate non-biodegradable waste, creating major waste disposal problems.

The use of plant-based products should help reduce these problems. Although burning fuel made from vegetable oil also produces carbon dioxide, this carbon dioxide has been removed from the atmosphere relatively recently when the crop that produced the oil was grown. The carbon dioxide released will be about the same quantity as the amount fixed, so there is no net change in carbon dioxide in the atmosphere, unlike when burning fossil fuels, which release carbon stores built up over long periods of time. The EU target is for 10% of our vehicle fuel consumption to be biofuels by 2020.

However, sustainability is not guaranteed by simply switching to products made from plants that were alive very recently rather than fossil plants. We need to consider the source of the plant product, and the energy used and pollution created during the production and transport of the product.

Using paper bags instead of plastic ones in supermarkets might seem like a good idea. But transport of the heavier bags to the stores uses more fuel and, if the bag gets wet, the shopping ends up on the ground as the bag disintegrates. A plant-based plastic bag that remains intact when wet, is light and will also biodegrade is a more sustainable alternative.

Does biodegradable packaging solve our waste problem? Most bacteria involved in degradation require oxygen, so what happens if the packaging is buried deep in a landfill site? The answer is remarkably little – newspapers have been retrieved from deep landfill after ten years and have still been readable! Even a frankfurter had not degraded. Closer to the surface, biodegradable packaging will degrade, though sometimes methane is generated which contributes to global warming.

 Did you know?

Brazil nuts and sustainability

The Brazil nut is the only nut that is normally harvested from the wild. Attempts have been made to grow Brazil nut trees in plantations, but their productivity is very low. This is probably due to the lack of orchid bee pollinators. Production is highest in undisturbed rainforest. Brazil nut production is important to the local economy, but there are concerns that in many areas the harvesting is not sustainable. A study found that in a sample of 23 populations which had been heavily harvested, young trees were missing from the populations. Over-exploitation of the nuts threatens the future of the trees and all the organisms that depend on the trees. Harvesting frequently disturbs the forest and affects productivity.

Although the long-term monetary value of the forest in terms of production of non-timber products is estimated to be greater than the value of the timber itself, there is widespread deforestation. The Brazil nut tree is the most highly protected tree in the rainforest, so frequently the surrounding trees are felled, leaving the Brazil nuts standing alone in an area used for agriculture. Although these trees can live for 200 years, the loss of the orchids and orchid bees mean the trees will not set seed and regenerate.

Is our planet big enough?

If moving to plant products sounds like a neat solution to our problems, consider the following difficulties.

We already cultivate a large proportion of the planet's surface for food, so there is not much room left for growing biofuels and other plant materials. Plants for biofuel are replacing rainforest (Figure 4.61) and also food crops.

Many of the remaining uncultivated areas are unsuitable for agriculture, and/or are essential for wildlife conservation or water conservation.

Intensive agriculture is itself very energy-demanding (think of ploughing, harvesting and any fertilising of the soil).

The human population is increasing rapidly.

Countries such as India and China are rapidly catching up with Western levels of consumption.

Some of the plant alternatives do not have the same desirable qualities as the oil-based products.

Some plant alternatives may require the consumption of as much or more energy in production and transport of the product, compared with the oil-based products.

Figure 4.61 Palm oil plantations are replacing tropical rainforests on a large scale with the loss of all the forest biodiversity. The area used for palm oil plantations has doubled in the last ten years in South East Asia, as demand in the West for biofuels and other products has increased. Palm oil is now found in one in ten supermarket products.

The richest one-sixth of the human population consumes approximately 80% of the world's resources. What is your own ecological footprint like? Your ecological footprint is the area of land you would require to support your current lifestyle. It assumes that the land you need is of average productivity for the Earth as a whole. It can then tell you how many planets like the Earth would be needed if everyone had an ecological footprint like yours. We need to consider how we can make changes to make our lifestyles sustainable.

> **Activity**
>
> Use the weblinks in **Activity 4.22** to work out your ecological footprint and sustainability. **A4.22S**

4.5 On the brink

The primary threat to most species and habitats is human activity. Threats include land development (which causes habitat destruction, fragmentation and degradation), over-exploitation, introduction of alien species, and pollution. Over the last 400 years, the extinctions of 484 animal species and 654 plant species have been recorded. This rate (which is probably an underestimate) is at least 70 times faster than expected from geological records.

The growing number of threatened species will not be saved from extinction merely by putting a fence around their remaining habitat and hoping for the best. As the golden lion tamarin story (Figure 4.62) illustrates, conservation of plants, animals or any other organisms requires a co-ordinated approach. Ideally, conservation management should be applied on site (*in situ*), protecting ecosystems and maintaining fragile habitats. In addition, vulnerable populations in the wild can be supported through the work of zoos and seedbanks. This is known as off-site (*ex situ*) conservation.

The role of zoos

The history of present-day zoos began in the 1750s, with the foundation of the first zoo in Europe by Emperor Franz Stephan in the grounds of Schönbrunn Palace just outside Vienna. London Zoo was founded in 1826, some six years before Charles Darwin set sail on the *Beagle*. Initially, the collection was privately studied by eminent scientists of the day, but opened its doors to the public in 1847. In those days before cameras and television, the public flocked to zoos, circuses and side shows to witness the wonders of the natural world. But almost nothing was known of these exotic animals' biology or behaviour, and little heed was paid to their needs in captivity. They were usually kept alone in bare cages designed for maximum visibility by the public (Figure 4.63).

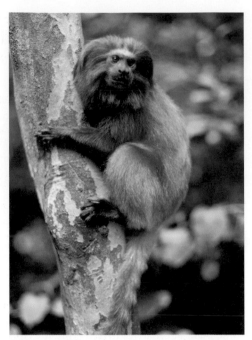

Figure 4.62 The golden lion tamarin (*Leontopithecus rosalia*) was on the edge of extinction in the early 1970s with only about 200 animals left in the wild. In 1974, the Golden Lion Tamarin Conservation Programme was set up. Scientists, conservationists and educators have worked together to protect and study the tamarin and its habitat so as to overcome the continuing threats to the species. Habitat management, captive breeding, reintroductions to the wild, education and research are all part of this international rescue effort to save the species. The golden lion tamarin is a conservation success story. Assessed as Critically Endangered in 2000, it has now been downgraded to Endangered as a result of over thirty years of conservation efforts. There are now about 1500 golden lion tamarins in the wild, with a further 475 in zoos involved in the conservation programme across the world.

Weblink

The International Union for the Conservation of Nature (IUCN) identifies over 11 000 species known to be threatened with extinction. You can find out exactly which species are threatened by visiting the IUCN Red List of Threatened Species website.

Figure 4.63 London's Royal Menagerie, 1812.

Today, zoos attract over 600 million visitors each year throughout the world. They manage over 1 million vertebrates in over 1000 registered collections and aim to have a significant impact in conservation, research and education.

Did you know?

Questioning the role of zoos

There is a growing awareness among those who run zoos of issues surrounding animal welfare, and the role zoos can play in saving species from extinction (Figure 4.64). As we learn more about habitat protection and the need to conserve endangered species in their natural environments, the conservation role of zoos, and the justification for keeping wild animals in captivity, is being questioned.

Some endangered species only exist in zoos. They are extinct in the wild. Few people argue that they would be better off extinct than living in captivity. Most zoos now try to have their animals living in naturalistic environments. UK zoos are regulated by the UK Zoo Licensing Act (1981), which aims to promote minimum standards of welfare, meaningful education, effective conservation, valuable research, and essential public safety. The introduction of the European Zoos Directive (1999/22/EC) by the European Union places a greater emphasis on conservation, education, research and welfare.

The Born Free Foundation believes wildlife belongs in the wild. It is dedicated to the conservation of rare species in their natural habitats, and the phasing out of traditional zoos. In 2000 it undertook a survey of the health status of UK zoos and published the report Zoo Health Check 2000.

The Born Free Foundation report expresses concerns about the welfare of animals in zoos, having observed many animals exhibiting stereotypic behaviours, such as repetitive pacing up and down and chewing of bars. They found that up to one in five collections that qualify as zoos under the current legislation do not have a zoo licence. Perhaps more importantly from the animals' perspective, 95% of species in zoos are not endangered, and are not part of European captive breeding programmes. The Born Free Foundation would rather see animals being bred in protected habitats in the wild.

Q4.32 What arguments might be presented for and against keeping animals in captivity in zoos?

Figure 4.64 Keeping elephants in captivity – should zoos take into account that in the wild most elephants live in family groups and range over very large areas?

Extension

Extension 4.6 allows you to consider some differing views on the role of zoos and debate the future of zoos. **X4.06S**

Centres for scientific research

One argument put forward for zoos is that they can play a vital role as research centres, enabling us to understand how to conserve particular species. The example below of mountain chickens at Jersey Zoo illustrates this.

How do mountain chickens breed?

The mountain chicken (*Leptodactylus fallax*) is not, as its name suggests, a bird, but is in fact one of the largest frogs in the world (Figure 4.65). It is found only on the Caribbean islands of Dominica and Montserrat. The mountain chicken's numbers have been declining because it has long been a national dish of the islands. It is hunted in large numbers for its meaty legs, which are used in traditional West Indian recipes. As their name suggests, their taste is somewhat like that of chicken.

Since 1995, over 75% of Montserrat has been engulfed by the continuing activity of the Soufrière Hills volcano. This volcano has destroyed much of the frog's remaining rainforest habitat, and left the already threatened species in a very precarious situation.

Little was known about the reproductive biology of *L. fallax*. The main reason for this was that it leads a secretive nocturnal life and so has proved difficult to study in the wild. If it was to be saved from extinction, more information was needed about its needs. In 1999, 13 frogs were taken to Jersey Zoo, allowing keeper scientists to find out more about their elusive lifestyle. Breeding trials have been set up to establish a captive breeding programme.

Q4.33 **a** Suggest why the mountain chicken's underground nest-building behaviour (Figure 4.66) has evolved.

b What would be some of the complications of conducting a study on these frogs in the wild?

Figure 4.65 *Leptodactylus fallax* – the chicken that's a frog. These frogs grow to 21 cm in length.

Figure 4.66 Spying on the foam nest. Unlike other frogs, mountain chickens lay their eggs in underground nests, covering them in foam.

Captive breeding programmes

Until recently, animal collections were unplanned and opportunistic. Zoos simply bought their animals from explorers and traders who would return from expeditions in ships laden with crates of anything they could catch and manage to bring back alive.

Today an important role of zoos is the successful breeding of the animals in their care. The aims of a captive-breeding programme include:

* increasing the number of individuals of the species if numbers are very low
* maintaining genetic diversity within the captive population
* reintroducing animals into the wild if possible.

There are European Endangered Species Programmes for over 200 species, with more than 400 institutions in Europe participating. Each species has an appointed co-ordinator/studbook-holder who advises on which animals should or should not breed and on the movement of animals between partner zoos, in order to maximise genetic diversity within the captive populations.

Weblink

To find out more about captive breeding programmes visit the European Association of Zoos and Aquaria website.

How genetic variation is lost

Genetic drift

In a small population, some of the alleles may not get passed on to offspring purely by chance, as shown in Figure 4.67. This change in the allele frequencies over time is known as genetic drift, and leads to a reduction in genetic variation.

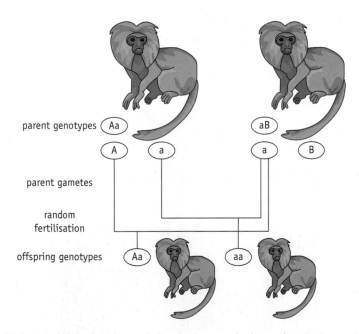

Figure 4.67 Genetic drift leads to less variation. In the example shown, the parents carry three alleles for a particular characteristic. By chance one of the alleles is not passed on.

> ### Key biological principle: Variation – the key to survival in a changing environment
>
> Genetic uniformity, where individuals within a population have similar genotypes, can be an advantage in a stable environment. But if the environment changes, a new disease emerges or a population moves, a genetically diverse population will be at an advantage.
>
> Scientists interested in the preservation of species hope that there is sufficient genetic diversity in the population to ensure that some of the individuals can cope with any new conditions, allowing the population to survive. This is natural selection and it results in adaptation, the accumulation of genotypes favoured by the environment. No natural environment remains unchanged forever. It is therefore essential for long-term survival that populations should be able to evolve as a result of natural selection. Such evolution is unlikely unless there is genetic variation to enable evolution to happen, or the mutation rate is very high.

Inbreeding depression

In a small population, whether in the wild or in captivity, the likelihood of closely related individuals mating increases. This inbreeding causes the frequency of homozygous genotypes to rise, with the loss of heterozygotes. Inbreeding results in individuals inheriting recessive alleles from both parents, and the accumulation of the homozygous recessive genotypes in the offspring. Many recessive alleles have harmful effects so **inbreeding depression** results. The offspring are less fit (less able to survive and reproduce). They may be smaller and not live as long, and females may produce fewer eggs.

Conserving genetic diversity

Conservation *in situ* (on site) to maintain the size of the wild populations is the best way to prevent genetic drift and inbreeding depression. However, *ex situ* (off-site) conservation can also play a role.

> ### Extension
>
> **Extension 4.7** allows you to discover some of the disadvantages and advantages of genetic uniformity in the 'ant supercolony story'.
> **X4.07S**

Keeping studbooks

The studbook for an individual species shows the history and location of all of the captive animals of that species in the places which are co-operating in an overall breeding plan. London Zoo, for example, keeps studbooks for many species including the slow loris, the Sumatran tiger (Figure 4.68) and the ocelot, and contributes to the studbook data for other species such as the golden lion tamarin.

Studbooks provide the raw data upon which all the breeding plans are based – the scientists' understanding of genetics shapes the breeding plans themselves. The conservation scientists must ensure that genes from all the founder members of the population (the original group of individuals, usually wild caught, on which the current population is based), or at least all remaining breeding adults, are retained and are *equally* represented in the subsequent generations (assuming

Figure 4.68 The Sumatran tiger is an endangered species which has been supported by a captive breeding programme in zoos. Zoo managers make decisions about breeding these animals based on their studbook data.

that these founders are unrelated). This requires that individuals who breed poorly in captivity must be *encouraged* to breed, whilst those that are particularly good breeders must be *limited* in their breeding success.

This approach, using a breeding plan, is a very different principle from that followed by zoos of fifty years ago, when they simply raised their captive populations from the best breeders. This seemed like common sense, but it obviously reduced genetic variation and began a process which made it less likely that there could be a successful reintroduction to the wild.

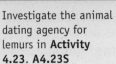

Activity

Investigate the animal dating agency for lemurs in **Activity 4.23. A4.23S**

Q4.34 Suggest which of the named animals in Figure 4.69 is genetically most valuable for breeding purposes.

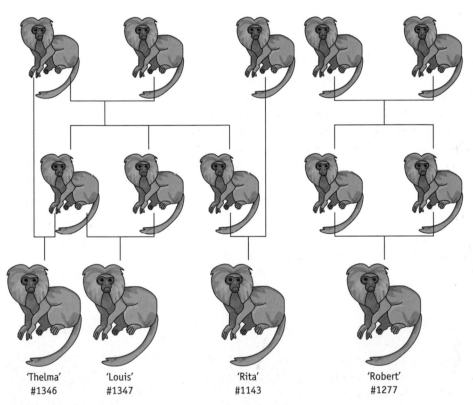

Figure 4.69 Some of the breeding undertaken as part of the golden lion tamarin programme.

More and more, studbook records are being supported by techniques of cytogenetics – looking at the structure of chromosomes – and of molecular biology – studying the nature of the genes themselves. These additional techniques are important because studbook data may be incomplete (since zoos in the past did not keep records as carefully as they do now). In addition, these new techniques can reveal whether some individuals are more closely related than is desirable for breeding purposes.

Reintroducing animals into the wild

One role that zoos have in species conservation is captive breeding for reintroduction. This involves breeding animals in captivity that are then returned to their native habitats. Clearly this can only work if the habitat is still intact.

Reintroduction is a complicated process, particularly when species need to learn new skills before they have the ability to survive in the wild. However, there are cases in which zoos have successfully reintroduced captive-bred animals into the wild, as the example below illustrates.

Going, going ... saved!

The Mauritius kestrel (*Falco punctatus*) was one of the rarest birds in the world, consisting of just four known individuals in 1974. The population was so small partly due to the extensive use of DDT as a pesticide to kill malaria-carrying mosquitoes. As the chemical worked its way up the food chain and accumulated, it was the kestrels as top predators which were most affected. Furthermore, the native habitat had become so degraded due to the removal of hardwood trees that it was not able to support a rapid increase in kestrel numbers without additional management techniques.

Carl Jones went out to Mauritius soon after 1974 and realised that the species could be saved. With the backing of the Durrell Wildlife Conservation Trust, whose headquarters are at Jersey Zoo, he set up a captive breeding centre on Mauritius where he took eggs from the wild birds and hatched them out in captivity (Figure 4.70). The young birds – fledglings – were then either returned to wild pairs of birds for rearing, or were 'hacked out'. 'Hacking out' involved them being taken out into the forest and gradually given more freedom and less food to encourage them to feed for themselves. Nest boxes were provided and natural cavities in trees modified to increase the number of suitable nest sites. In addition, predators, especially rats, were controlled at release and nest sites. As a result, there are now over 800 Mauritius kestrels slicing through the forest air.

Figure 4.70 Young captive-bred Mauritius kestrel, *Falco punctatus*.

Q4.35 What would be both the advantages and disadvantages of keeping the young Mauritius kestrels in open nest boxes in the forest?

Activity

In **Activity 4.24** you analyse the reintroduction programmes for the Mauritius kestrel and ruffed lemurs. **A4.24S**

Checkpoint

4.7 Summarise the roles of zoos in the conservation of endangered species.

The Millennium Seed Bank (MSB)

Plants are threatened worldwide by habitat destruction, climate change and over-harvesting. Protecting and managing habitats conserves plants *in situ*, but they remain at risk from man-made and natural disasters so there is also a role for *ex situ* conservation, using a variety of methods including seed banks and botanic gardens. The aim of the Millennium Seed Bank Project is to conserve seed samples from threatened species of plants, with 10 000 species already banked. Seeds are collected around the world. Some are kept in their countries of origin, and some are sent to the Millennium Seed Bank in the UK (Figure 4.71).

Figure 4.71 The Millennium Seed Bank at Wakehurst Place in Sussex. The Millennium Seed Bank Project, a worldwide seed conservation network, aims to safeguard targeted wild plant species.

Most plants produce large numbers of seeds, so collecting small samples is unlikely to damage a wild population. Most seeds are small and easy to store, and can survive in a desiccated state for many years. But how do we know how long the seeds will live? What can we do to enable them to survive for a longer time?

The technology of seed preservation is improving all the time as more research is carried out. Seeds survive longer if kept dry and cool (Table 4.6). For every 1% reduction in seed moisture content, seed life span doubles, and for every 5 °C reduction in temperature, seed life span doubles. Once the seeds' identification has been verified and they have been cleaned and dried, the seeds are stored at -20 °C.

Seed	Longevity (years)
Sunflower	165
Lettuce	447
Oilseed rape	843
Soybean	1122
Wheat	5079
Sugar beet	10 542

Table 4.6 Predictions of longevity under typical seed bank storage conditions.

About a month after the seeds have been placed in cold storage, a sample is taken out and germinated on agar plates, to ensure that the seeds survive in the storage conditions. Germination is then tested about every 10 years to check that they are still alive. If germination falls below 75%, then seeds will be grown to collect a new seed sample, which is then placed back into storage.

Some of the tests on seed viability are now being carried out as school projects, supported by The Royal Botanic Gardens, Kew (Figure 4.72).

Figure 4.72 Schools have tested the effect of storage time on germination in a number of species as part of the Save Our Seeds project.

Activity

Use **Activity 4.25** to find out more about the Millennium Seed Bank project. **A4.25S**

The Millennium Seed Bank Project aims to collect seed samples from 10% of the world's seed plants by 2010. They are focusing their efforts on plants living in drylands, as these are some of the most threatened, and potentially useful to humans.

Q4.36 Why are plants from drylands likely to be threatened?

Q4.37 Why are dryland plants likely to be useful to humans?

The seed collections are also used for research, habitat restoration and species reintroductions. Plants of *Silene tomentosa,* a species from Gibraltar thought to be extinct in the wild, were grown from seeds stored in the MSB and successfully reintroduced to Gibraltar. Seeds from UK native species such as *Damasonium alisma* (starfruit), *Corrigiola litoralis.* (strapwort) and *Apium repens* (creeping marshwort) have also been used for reintroductions.

Checkpoint

4.8 So why are there so many different species? Use the relevant principles in this topic to answer this question.

Activity

Use **Activity 4.26** to check your notes at the end of this topic. **A4.26S**

Topic test

Now that you have finished Topic 4, complete the end-of-topic test. Congratulations – you have completed the AS!

Answers to in-text questions
Topic 1

Q1.1 Movement of oxygen; carbon dioxide; and other products carried by blood relies on diffusion in animals with an open circulatory system; diffusion is only fast enough for small organisms;

Q1.2 Blood can pass slowly through the region where gaseous exchange takes place; maximising the transfer of oxygen and carbon dioxide; and then be pumped vigorously round the rest of the body; enabling the organism to be very active;

Q1.3

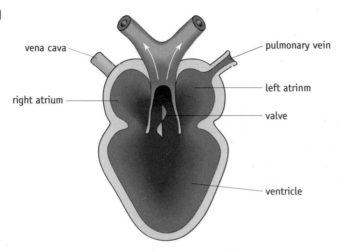

vena cava

pulmonary vein

left atrinm

right atrium

valve

ventricle

Q1.4 Some mixing of oxygenated and deoxygenated blood; in the ventricle;

Q1.5 Thick layer of mainly elastic fibres; to allow expansion and recoil of artery; surrounded by thick layer of mainly collagen fibres; tough and durable to withstand high pressure;

Q1.6 The blood pressure in the left ventricle falls below that in the aorta; leading to the closure of the semilunar valve between the left ventricle and the aorta; the blood pressure in the right ventricle falls below that in the pulmonary artery; leading to the closure of the semilunar valve between the right ventricle and the pulmonary artery;

Q1.7

	Probability	
	Percentage	Per million people
Heart disease	0.23%	2300
Lung cancer	0.06%	600
Road accidents	0.006%	60
Accidental poisoning	0.00017%	17
Injury purposely inflicted by another	0.00049%	4.9
Railway accidents	0.00007%	0.7
Lightning	nearly 0	0.1

Q1.8 Ten students out of 1300 $\left(\frac{10}{1300}\right)$ is less than 1%; assumptions: the same exposure for each student; e.g. the same frequency of swimming for each student; the same length of each term with the same frequency of swimming each term; and the same viral load in and around the pool each term;

Q1.9 Yes; risk of catching chlamydia in London in 2005 is 21 215 in 7 500 000

or 1 in $\dfrac{7\,500\,000}{21\,215}$

= 1 in 354 or 0.28%

Risk of catching chlamydia elsewhere in the UK in 2005 is 88 743 in 52 700 000

or 1 in $\dfrac{52\,700\,000}{88\,743}$

= 1 in 594 or 0.17%

Q1.10 **a** no causal link; hot weather increases the number of ice cream sales and the number of people going swimming in shark-infested waters;

b no causal link; both these variables increase with age;

c causal link; smoking greatly increases the chance of developing lung cancer because tars and other substances in the smoke damage cells in the lungs;

Q1.11 No; the study only shows a correlation between smoking and lung cancer; evidence of how smoke damages living cells showed that smoking does cause cancer;

Q1.12 It would not be valid to extrapolate these results to non-white populations because the cohort is not representative of the general population which has a wide range of ethnic and racial groups; further studies have shown that the same results are true for most ethnic groups;

Q1.13 It increases; greatly;

Q1.14 Not necessarily; it might be that behaviours early in life greatly affect one's subsequent chances of developing cardiovascular disease;

Q1.15 No; before the age of 74 males are more likely to die from cardiovascular disease than are females; after this, the greater number of deaths among women is simply because they greatly outnumber men;

Q1.16 The prevalence data in Table 1.1 is likely to be less reliable than the mortality data in Table 1.2; some people who have CHD may not recognise and report the disease;

Q1.17 The data provide some support for the view that until the menopause a woman's reproductive hormones offer her protection from coronary heart disease in that deaths from cardiovascular diseases increase more steeply among women over the age of 50 than they do among men; quantify; but there are other possible explanations for this; so it would be premature to draw such a conclusion with any confidence;

Q1.18 Blood pressure rises in the lungs due to a back-up of blood in the pulmonary capillaries; causing tissue fluid to build up within the lungs; impairing gas exchange;

Q1.19 **a** Approximately 478 000 calories;

b Approximately 478 Calories;

Q1.20 Carbon, hydrogen and oxygen are found in a 1:2:1 ratio;

Q1.21 In galactose the –OH groups on carbon 1 and carbon 4 lie above the ring whereas in glucose the –OH groups on carbon 1 and carbon 4 lie below the ring;

Q1.22   **a** sucrose

glucose   fructose

b maltose

glucose   glucose

c lactose

galactose   glucose

Q1.23

Q1.24   Energy requirements tend to increase up to the age of 15–18 years; after the age of 18 years energy requirements decrease; males' requirements remain static between 19 and 59 and then decrease with age; females' requirements decrease slightly in their 50s and continue to decrease with age; the decrease with age is partly due to a reduction in the basal metabolic rate (BMR) and to a reduced level of activity; on average, males have slightly higher requirements than females throughout life;

Q1.25   If people are more active, larger physical activity level factors would need to be used (1.9 is used for very active adults);

Q1.26   30.1, moderately obese;

Q1.27   Waist to hip ratio 0.79; he is at low risk of heart disease;

Q1.28   There is a positive correlation between increasing serum (blood) cholesterol and higher death rate for 30-39 year olds and 40-49 year olds; overall there is also an upward trend for 50-59 year olds but the value for 6.21 is lower than for 5.69 serum cholesterol;

Q1.29   The fact that pre-menopausal women generally have higher HDL:LDL ratios than men would be expected to lead to their having lower rates of coronary heart disease;

Q1.30 LDL levels in the blood will fall; less saturated fat triglycerides in the blood to combine with cholesterol and protein to form LDLs;

Q1.31 Six phenotypes; 3/3, 4/3, 3/2, 4/4, 4/2, 2/2;

Q1.32 No increase or decrease in CVD risk compared to 'normal';

Q1.33 Antioxidants protect against radical damage; radicals are highly reactive; and can damage many cell components; they have been implicated in the development of heart disease and some other diseases;

Q1.34 People underestimate the risk associated with high cholesterol; they are unwilling and find it difficult to make the lifestyle changes needed to lower cholesterol;

Q1.35 People who have had a heart attack may be more motivated to follow a strict diet; they probably started with higher blood cholesterol levels making it easier for them to reduce these; they may live in institutions where control over diet is greater;

Q1.36 **a** To decide on the appropriate drug treatment her risk factors for heart disease should be considered; if she had high blood pressure, antihypertensive drugs might be prescribed to reduce her blood pressure; if she had high cholesterol, a drug to reduce blood cholesterol levels would be used; if there was a high risk of clotting then anticoagulant or platelet inhibitory drug treatment would be given; in deciding which particular drug to prescribe the benefits of the treatment would be considered along with any risk of side effects;

b No, it would not be sensible; for some people the risk of internal bleeding with aspirin may outweigh the benefit of reduced risk of cancer, particularly if their risk of bowel cancer was unknown or very low;

Q1.37 Cholesterol levels in the blood at the time of the stroke; blood pressure at the time of the stroke; report of the state of the artery when surgeons operated; genetic screening for the alleles that increase susceptibility to CVD;

Q1.38 Eat a healthy diet; a diet low in saturated fats, low in salt, high in antioxidants; take regular exercise; do not smoke; avoid stress;

Topic 2

Q2.1 Pathogenic microorganisms have time to multiply, resulting in illness or infection;

Q2.2 Acid in the stomach kills the microorganisms;

Q2.3 For A, SA = 6, Vol = 1, SA:Vol = 6; for B, SA = 24, Vol = 8, SA:Vol = 3; for C, SA = 96, Vol = 64, SA:Vol = 1.5;

Q2.4 **a** Its surface area increases by a factor of 4;

b Its volume increases by a factor of 8;

c Its surface area to volume ratio halves;

Q2.5 The surface area to volume ratio would continue to fall; the organism would not be able to exchange enough substances to survive;

Q2.6 Hippopotamus;

Q2.7 For D, SA = 34, Vol = 8, SA:Vol = 4.25; for E, SA = 28, Vol = 8, SA:Vol = 3.5;

Q2.8 Volumes are all the same, 8; but the more elongated the block the greater the surface area and thus the larger the surface area to volume ratio;

Q2.9 Tapeworm;

Q2.10 Desiccation/dehydration problems; surface also has protective function;

Q2.11 I;

Q2.12 Kidney; capillaries;

Q2.13 They are carried in the bloodstream;

Q2.14 Slower mixing at lower temperatures supports the model because movement of molecules in fluids is slower at lower temperatures;

Q2.15 The kinks in the fatty acids prevent them lying very close together; this creates more space in which the molecules can move;

Q2.16 **a** Diffusion;

b Active transport;

c Active transport;

d Facilitated diffusion / channel protein;

e Active transport;

f Osmosis;

Q2.17 Salt is normally reabsorbed from sweat using the CFTR channel; with CF this does not function; so the salt is not absorbed making saltier sweat;

Q2.18 It is a polymer; of nucleotides;

Q2.19 A and G both have a two-ring structure whereas C and T have only one ring; the bases pair so that there are effectively three rings at each of the 'rungs' of the DNA molecule; making the molecule a uniform width along its whole length; the number of bonds that form between the molecules determines which bases pair; only two hydrogen bonds form between A and T, whereas three bonds form between C and G;

Q2.20 T A G G G A C T C C A G T C A;

Q2.21 **a** Messenger RNA;

b Transfer RNA;

Q2.22 **a** TCA;

b AGU;

c UCA;

Q2.23 A U G U A C C U A A G G C U A

Q2.24 5;

Q2.25 UAC; AUG; GAU; UCC; GAU;

Q2.26 Conservative replication: light and heavy bands at top and bottom of tube; fragmentary and semi-conservative replication: medium band in middle of tube;

Q2.27 a 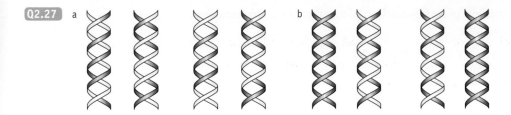 b

Q2.28 If the fragmentary model was correct the DNA would all be medium density and appear as a band in the middle of the tube;

Q2.29 GTCAGTCCG;

Q2.30 **a** The sixth base is G rather than T;

b The second base is A instead of T;

c The fifth and sixth bases are inverted;

d Omission of the fourth base, A;

e An additional base, G, is added after the fifth base;

Q2.31 In situations one and two both parents are probably carriers and there is a 1 in 4 chance that any child they have has CF; if the father in situation three is not a carrier then none of the woman's children will get the disease; they will all be carriers receiving a defective allele from her and a 'normal' allele from their father; if these children were to have children with a carrier there would be a 1 in 4 chance that each child would have CF;

Q2.32 James and/or Margaret must be a carrier of the CF gene mutation; Polly has inherited the recessive allele from one of her parents; both Polly and Wilf are CF carriers and have passed the recessive alleles on to Frank;

Q2.33 The test will not be completely reliable (it is only about 80–85% sensitive); there should not be false positives but there will be false negatives where an individual has a CF mutation but does not have one of the mutations which the test can detect;

Q2.34 Yes; if the parents believe for religious or other reasons that abortion is wrong; if they consider that the risk of miscarriage is too high; if they want to have the baby even if it will have cystic fibrosis;

Q2.35 Have a child accepting the 1 in 4 risk of it inheriting CF; use artificial insemination by donor to avoid the risk; use *in vitro* fertilisation and have the embryo screened before implantation; use prenatal screening and decide whether to continue with the pregnancy if the foetus has the disease;

Topic 3

Q3.1 Cell surface membrane; smooth ER; mitochondrion; smooth ER; rough ER; nuclear membrane; nucleolus; nuclear membrane; rough ER; Golgi apparatus; cell surface membrane;

Q3.2 Lysosome; free ribosomes;

Q3.3 A – nucleolus; B – nuclear pore; C – nuclear membrane; D – mitochondrion; E – smooth endoplasmic reticulum;

Q3.4 XYZ; **XYZ**; X**Y**Z; XY**Z**; **XYZ**; X**YZ**; **XY**Z; **XYZ**;

Q3.5 16 (= 2^4); 32 (= 2^5);

Q3.6

A   B

Q3.7 G1 = 8 hours; S = 4 hours; G2 = 8 hours; division = 4 hours;

Q3.8 No distinct chromosomes are visible; protein synthesis is taking place;

Q3.9 Ribosomes; endoplasmic reticulum;

Q3.10 Condensing allows the DNA molecules to move around the cell without getting tangled up;

Q3.11 The pores in the nuclear envelope are not large enough for whole molecules of DNA to pass through; once it disintegrates the chromosomes can move freely through the cell;

Q3.12 Mitosis is a continuous process but a slide or photograph is a static image; the shorter the stage, the less likely a cell is to be in that stage when the action is 'stopped' as the slide is made or the photograph taken; it can be deduced that the fewer cells seen in any one stage, the quicker that stage is; anaphase is shorter than telophase because there is only one cell in anaphase and three in telophase; a larger sample would be needed to give a more accurate timing;

Q3.13 **a** Asexual; sexual;

b Sexual; asexual;

c Sexual; asexual;

d Sexual; asexual;

e Asexual; sexual;

f Asexual; sexual;

g Asexual; sexual;

Q3.14 G1; G2;

Q3.15 The adult cell providing the genetic information to create Dolly was a specialised mammary gland cell; the successful birth of Dolly suggests that the cell must have contained all the information for making a complete organism;

Q3.16 **a** Tissue; organ; organ; cell;

b Transport; photosynthesis; excretion; communication; or similar;

Q3.17   **a** Sepals and petals;

b

Q3.18

	Bbb	bbb
Bbb	BB bb bb (2)	Bb bb bb (1)
BbB	BB bb Bb (3)	Bb bb Bb (2)
bbb	Bb bb bb (1)	bb bb bb (0)
bbB	Bb bb Bb (2)	bb bb Bb (1)

The number of alleles adding pigment is shown in brackets; the child with three of these alleles will have light brown eyes; there is a 1 in 8 chance of the couple having a child with brown eyes;

Q3.19   **a**

b Height shows continuous variation;

c i Parental phenotypes   Baseline height plus 30 cm
          Possible gametes   AB Ab aB ab

ii

	AB	Ab	aB	ab
AB	AABB 40	AABb 35	AaBB 35	AaBb 30
Ab	AABb 35	AAbb 30	AaBb 30	Aabb 25
aB	AaBB 35	AaBb 30	aaBB 30	aaBb 25
ab	AaBb 30	Aabb 25	aaBb 25	aabb 20

A and B contribute 10 cm above baseline height while a and b add 5 cm to baseline height; the ratio of heights above baseline (cm) 20, 25, 30, 35, 40 in the offspring is 1:4:6:4:1;

Q3.20   1 and 2 are the result of genotype; 3 to 6 are environmental;

Q3.21   MSH activates melanocytes by attaching to MSH receptors on the cell surface; without these receptors, MSH has no effect and no melanin is made in the hair follicles;

Q3.22 In humans sunlight stimulates MSH receptor production; there is more sunlight in the summer;

Q3.23 The tips are slightly cooler and so the enzyme remains active; melanin is made and tips are darker;

Q3.24 Humans with this condition have heat-sensitive tyrosinase (like Himalayan rabbits and Siamese cats); so in the warmer parts of the body, such as the armpits, tyrosinase does not function and the hair is white; on the cooler surfaces of the body, tyrosinase works and the hair is dark;

Q3.25 Each time DNA replicates before mitosis, a few errors will be made (mutations); the higher the rate of mitosis the more likely it is that mutations will occur in genes controlling cell activity, which can result in cancer;

Q3.26 When the embryo grows into an adult, by mitosis, the cells giving rise to the ovaries or testes may have the DNA error; thus gametes with faulty DNA could form; the cancer-causing error could be passed on in these gametes to the next generation;

Q3.27 If the single normal allele becomes damaged, a person with only one normal allele would no longer produce p53; whereas if someone with two normal alleles gets one of them damaged, they would still make p53 and be protected from cancer;

Q3.28 If chemotherapy and radiotherapy work by activating the genes for p53; they will have no effect if the genes controlling p53 synthesis are not working;

Topic 4

Q4.1 **a** A is a different species from B since 2% is a high percentage of genetic difference;

b A and C are either the same species or very closely related species as they can interbreed and produce fertile young;

c B and D are closely related but distinct species as the males that result from interbreeding are infertile;

d E and F are the same species since there is no genetic difference between them;

Q4.2 **a** Ensures that pollen comes only from a member of its own species;

b If the insect is not present (or becomes extinct) the plant will not be pollinated and could not reproduce;

Q4.3 The woodpecker might be better adapted and displace the finch, or the woodpecker would find the niche already filled and not be able to survive;

Q4.4 **a** Sight of the red spot on beak; chick obtains food from parent;

b Sight or sound of predator; draws predator away from young in the nest;

c Humidity of surroundings; woodlice will be less likely to dry out in damper conditions; in drier conditions they move around more so they are more likely to move to damper conditions; once in damp conditions they move around less so they stay in the more humid conditions;

Q4.5 Active transport;

Q4.6 Bumblebees with longer tongues tend to take nectar from flowers with longer corollas; this reduces competition between the different species; however, there is some niche overlap; bramble and white clover were used by a wide range of species;

Q4.7 Brazil nut – hard coat to fruit; structural;

Agouti – sharp teeth; structural;

Euglossine bee – long specialised tongue; structural;

Orchid – bee attractant chemical produced; physiological;

Q4.8 Ants gain protection; shelter;

Q4.9 No; because there might be some disadvantages in having too long a tongue, for example it might be more likely to break or be difficult to carry in flight;

Q4.10 1A; 2B; 3C; 4D; 5E; (3B and 2C are also acceptable;)

Q4.11 A genetic mutation; this might work, for example, by making the mosquito's exoskeleton impermeable to the insecticide, or it might code for an enzyme which breaks down the insecticide;

Q4.12 Birds of prey are slow breeding and have a relatively small population/gene pool;

Q4.13 With a larger gene pool there are more likely to be individuals which are able to survive if conditions change; these survivors could give rise to a population which is adapted to the new conditions;

Q4.14 **a** *Ranunculus*;

b Meadow sweet and dropwort; they both belong to the same genus;

Q4.15 *Chaetodon auriga*;

Q4.16 They have a long extended part to their dorsal fins;

Q4.17 Both woods have the same species richness, with eight species of tree; but the ancient wood has more evenness, i.e. the numbers of each species are more similar; the ancient wood is considered by ecologists to have greater biodiversity, despite (in our example) having the same number of tree species;

Q4.18 Plants are easier to identify; a much greater proportion of the world's plants than animals have been identified; plant diversity is related to animal diversity;

Q4.19 Hotspots appear to be in the warmer regions of the world; greater availability of sunlight and precipitation provide the best conditions for plant growth; plant and associated animal diversity will tend to be lower where conditions are less favourable; areas more recently glaciated will have lower diversity;

Q4.20 **a** Nucleus;

b Chloroplasts;

c Golgi apparatus;

d Cell membrane;

e Vacuole;

Q4.21 One of the molecules will have to be inverted; so that the two –OH groups lie alongside each other and can react;

Q4.22 It links the –OH group on the first carbon of one glucose with the –OH attached to the fourth carbon of another glucose;

Q4.23

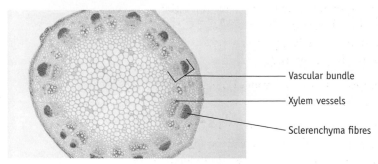

Vascular bundle

Xylem vessels

Sclerenchyma fibres

Q4.24 Release of enzymes from within the vacuole may digest cell contents;

Q4.25 The specific heat capacity of water is the amount of energy in joules required to raise the temperature of 1 cm^3 (1 g) of water by 1 °C; it is very high for water because a large amount of energy is required to break the hydrogen bonds between water molecules;

Q4.26 Active transport;

Q4.27 Strength suited to use; length; ease and reasonable cost of extraction; durability; resistance to decay;

Q4.28 These parts come into greater contact with the soil; so are at greatest risk of microbial infection;

Q4.29 He tested the drug on patients with symptoms of the disease; recorded any side effects; used standard procedure to discover the effective dosage; slowly increased dose until patients experienced diarrhoea and vomiting and then reduced dose slightly; recorded all results meticulously and published results;

Q4.30 **a** It is difficult for those involved in the production of the new compound to be completely objective about it;

b Phase III is more rigorous; larger sample size; double-blind; use of statistics;

c Yes; because it is not known for certain if the new compound is effective; patients freely consent to participate in trials;

d The influence of mind over the body; the patient may have been getting better anyway due either to chance or their own immune response;

Q4.31 **a** wind;

b wind;

c self-dispersal;

d animals;

e animals;

f water;

Q4.32 *Arguments for*: role in academic research; captive breeding with careful use of studbooks to maintain genetic diversity; reintroduction programmes; education;

Arguments against: animals are exhibited on the basis of their 'crowd pulling' power, rather than on their endangered status; animals are kept in inappropriate conditions; both in terms of their physical environment (poor floors in cages, for example); and their mental and social well-being (limited feeding stimuli or atypical social grouping); capture of animals for exhibition may seriously deplete wild populations; reintroduction of species to the wild cannot be guaranteed;

Q4.33 **a** Scarcity of surface water; competition for these water bodies; high densities of aquatic predators; foam reduces risk of desiccation;

b Financial cost of flying someone out, their food and accommodation; logistics of working in the field raise problems such as finding the nests in the first place; avoiding disturbance to egg-laying and care of young; the impact of the weather on conducting regular observations; liaison with local people;

Q4.34 Robert; he is the only offspring from the two founders on the right; his parents are related so he is inbred but he can be paired with an unrelated animal;

Q4.35 *Advantages*: same environment as they will have to get used to after reintroduction; *Disadvantages*: risks of disease and death due to weather; parasites; predators;

Q4.36 Drylands are very sensitive to intensive grazing, soil erosion and droughts;

Q4.37 Many people live in marginal lands; wild plants may provide useful alternative crops or be wild relatives with useful genes which could be used in plant breeding.

Index

Picture credits

The publisher would like to thank the following for their kind permission to reproduce their photographs:

(Key: b-bottom: c-centre: 1-left: r-right: t-top)

Alamy Images; Danita Delimont / Adam Jones l94bl; Edward Parker l42bl, 169b; Food Drink and Diet / Mark Sykes 51b; William Stuart Robinson 166; **Angela Hall**: 178t, 189; **Ardea**: Bill Coster 148; lan Beames 159b1; Ron & Valerie Taylor 159br, 159c, 159cr Valerie Taylor 158, 159c1; **Biophoto Associates**; 103, 177cr; **John Birdsail Social Issues Photo Library**: John Birdsall 19; **Gareth Boden**: 35b; **CEC / Mark M Boulton**; 741t; **Trevor Clifford**: 35t; **Corbis**: 2; Historical Picture Archive 192b1; Peter Johnson 164; **Dennis Kunkel Microscope Inc**; 64; **DK Images**: 168; **Durrell Wildlife Conservation Trust / Gerardo Garcia**; 194br; **FLPA Images of Nature**; Jurgen & Christine Sohns 100br. 196; M. Szadzuik and R. Zinck 184; Nigel Cattlin 181c; Peter Entwistle 145cr; R. P. Lawrence 193br; Winfried Wisniewski 193; **Getty Images**: Bob Elsdale 102t; Mike Timo 3; Photodisc 54. 114t; **iStockphoto**; Anzeletti 48; **National Himalayan Rabbit Club**; 139; **Nature Picture Library**: Dave Watts 145tc; Georgette Douwma 119b1; Nick Garbuitt 198; Philippe Clement 145t1, 147; **NHPA Ltd / Photoshot Holdings**: Cede Prudente 191; Ken Griffiths 144; Stephen Dalton 157c; **Peter Gould**; 29; **Royal Botanic Garden Kew**: 199; **Royal Geographical Society**: Ranulph Fiennes 4l; **Science Photo Library Ltd**: 13c, 56, 185b; Alex Bartel 170; Andrew Syred 125bc, 177c1; Andy Walker, Midland Fertility Services 100tr, 120; Antonia Reeve 53; Biophoto Associates 10t, 141, 60; BSIP, Laurent 137; BSIP, Laurent H. American 46; BSIP / SERCOMI 102tc; Carlos Munoz-Yague / Eurelios 183c; Claude Nuridsany & Marie Perennou 181b; CNRI 125bl; D. Phillips 107; David Scharf 152; Don W. Fawcett 67; Dr Gopal Murti 116c; Dr Yorgas Nikas 15; Dr. Don Fawcett 105; Dr. Jeremy Burgess 113, 118b, 132b, 155, 163, 171, 175, 185c, 187; Dr. Keith Wheeler 157t; Dr. Morley Read 142br; Elena Kiseleva 85; Eye of Science 89, 132cl, 132cr; GCa 17; Geoff Kidd 121t; George Bernard 119br; GJLP 13b; Hattie Young 93; James Stevenson 141; John Eastcott & Yva Momatiuk 138b1; John Radcliffe Hospital 9; Kenneth Eward / Biopgrafx 77; James King-Holmes 122, 123; Manfred Kage 119t; Mark Thomas 43; Pascal Goetgheluck 97, 100tl; Photo Researchers 81; Pr, G. Gimenez-Martin 114b, 116b, 116t, 117b, 117c, 117t; Prof. P. Motta / Dept. of Anatomy / University "La Sapienza"rome 57; R. Maisonneuve, Publiphoto Diffusion 177b; Rosenfeld Images Ltd 33; Rosenfeld Images Ltd. 121b; Saturn Stills 26, 97; Shelia Terry 51c, 174; Sinclair Stammers 119bc; Steve Gschnieissner 14r, 100bl, 125br, 125t; The Biocomposites Centre / Eurelios 182; Tom McHugh 192t; Vilker Steger 183b; William Ervin 138br, 142bc; **Anne Scott**: 130, 150, 169t1, 169tr, 200; Mark Tolley 4b, 4t; Peter Kempson 5c, 5t; **Tim Gainey / Photographers Direct**: 146

All other images © Pearson Education

We are grateful to the following for permission to reproduce copyright material:

British Heart Foundation for figure 1.7 on page 6 'Deaths from cardiovascular disease' reproduced with the kind permission of the British Heart Foundation; Office for National Statistics for the tables 1.1 and 1.2 on page 25 "Rates per 1000 population reporting longstanding diseases of the circulatory system by sex and age, 2004, Great Britain" from *General Household Survey 2004* and "Mortality data from diseases of the circulatory system for England and Wales in 2004" from 2004 registrations © Crown copyright 2005; Elsevier Limited for figure 1.45 on page 41 from the article "Obesity and the risk of myocardial infarction in 27000 participants from 52 countries: a case-control study" by Yusuf Salim, Hawken Steven, Ôunpuu Stephanie, Bautista Leonelo, Franzosi Maria Grazia, Commerford Patrick, Lang Chim C, Rumboldt Zvonko, Onen Churchill L, Lisheng Liu, Tanomsup Supachai, Wangai, Jr Paul, Razak Fahad, Sharma Arya M and Anand Sonia S published in *The Lancet* 5-11 November 2005, Volume 366, Issue 9497, pgs 1640-1649 copyright © Elsevier Limited, reprinted with permission; Nottingham University Press for figure 1.47 on page 42 from *Diet, Lipoproteins and Coronary Heart Disease: A Biochemical Perspective* by Heather Mangiapane and Andy Salter published by Nottingham University Press 1999 copyright © Nottingham University Press, reprinted with permission; Pearson Education Inc, for figure 2.5 on page 61, figure 3.36 on page 127 and figure 4.41 on page 176 from *Biology* 5/ed by Neil A. Campbell, Jane B. Reece and Lawrence G. Mitchell copyright © 1999 by Benjamin/Cummings, an imprint of Addison Wesley Longman, Inc. Reprinted by permission of Pearson Education, Inc; PFD for an extract on page 81 from "The Play: Lifestory" by William Nicholson copyright © William Nicholson. Reproduced by permission of PFD (www.pfd.co.uk) on behalf of William Nicholson; Penguin Group (UK) for figure 4.2 on page 144 adapted from of 'Number of Living Species of All Kinds of Organisms Currently Known' from *The Diversity of Life* by Edward O. Wilson (Allen Lane, The Penguin Press 1993, Penguin Books 1993, reprinted 2001) © copyright Edward O Wilson, 1992, 2001, reprinted with permission of Penguin Books; Bob Ellis and Gillian Beckett for figure 4.9 map on page 148 "The distribution of Danish Scurvy grass, *Cochlerea danica* in Norfolk" published in *A Flora of Norfolk* by Gillian Beckett and Alec Bull & Robin Stevenson 1999, reprinted with permission of Bob Ellis and Gillian Beckett; Curriculum Research & Development Group for figure 4.17 on page 158 "Key to butterfly fish" from *HMSS The Living Ocean* pages 14-21 by E.B. Klemm, S.A. Reed, F.M. Pottenger, C. Porter and T.W. Speitel 1995 copyright © CRDG, University of Hawai'i at Manoa reproduced by permission of the Curriculum Research & Development Group; America Society of Microbiology and Carl Woese, Professor of Microbiology at University of Illinois for figure 4.23 on page 163 "There must be a prokaryote somewhere: microbiology's search for itself" by C R Woese published in *Microbiology Review* March 1994, 58(1): 1–9 copyright © American Society of Microbiology, Journals Department; and GLTCP for figure 4.67 on page 195 and figure 4.69 on page 197 adapted from http://nationalzoo.si.edu/ConservationAndScience/EndangeredSpecies/GLTProgram/ZooLife/Science/

Every effort has been made to trace the copyright holders and we apologise in advance for any unintentional omissions. We would be pleased to insert the appropriate acknowledgement in any subsequent edition of this publication.